Stieltjes Differential Calculus with Applications

TRENDS IN ABSTRACT AND APPLIED ANALYSIS
ISSN: 2424-8746

Series Editor: John R. Graef
   The University of Tennessee at Chattanooga, USA

This series will provide state of the art results and applications on current topics in the broad area of Mathematical Analysis. Of a more focused nature than what is usually found in standard textbooks, these volumes will provide researchers and graduate students a path to the research frontiers in an easily accessible manner. In addition to being useful for individual study, they will also be appropriate for use in graduate and advanced undergraduate courses and research seminars. The volumes in this series will not only be of interest to mathematicians but also to scientists in other areas. For more information, please go to http://www.worldscientific.com/series/taaa

Published

Vol. 12   *Stieltjes Differential Calculus with Applications*
         by Svetlin G. Georgiev & Sanket Tikare

Vol. 11   *Stochastic versus Deterministic Systems of Iterative Processes*
         by G. S. Ladde & M. Sambandham

Vol. 10   *Nonlinear Higher Order Differential and Integral Coupled Systems: Impulsive and Integral Equations on Bounded and Unbounded Domains*
         by Feliz Manuel Minhós & Robert de Sousa

Vol. 9   *Boundary Value Problems for Fractional Differential Equations and Systems*
         by Bashir Ahmad, Johnny Henderson & Rodica Luca

Vol. 8   *Ordinary Differential Equations and Boundary Value Problems Volume II: Boundary Value Problems*
         by John R. Graef, Johnny Henderson, Lingju Kong & Xueyan Sherry Liu

Vol. 7   *Ordinary Differential Equations and Boundary Value Problems Volume I: Advanced Ordinary Differential Equations*
         by John R. Graef, Johnny Henderson, Lingju Kong & Xueyan Sherry Liu

More information on this series can be found at https://www.worldscientific.com/series/taaa

Trends in Abstract
and Applied Analysis – Volume 12

Stieltjes Differential Calculus with Applications

Svetlin G Georgiev
Sorbonne University, France

Sanket Tikare
Ramniranjan Jhunjhunwala College, India

World Scientific

NEW JERSEY • LONDON • SINGAPORE • BEIJING • SHANGHAI • HONG KONG • TAIPEI • CHENNAI

Published by

World Scientific Publishing Co. Pte. Ltd.
5 Toh Tuck Link, Singapore 596224
USA office: 27 Warren Street, Suite 401-402, Hackensack, NJ 07601
UK office: 57 Shelton Street, Covent Garden, London WC2H 9HE

Library of Congress Control Number: 2024047954

British Library Cataloguing-in-Publication Data
A catalogue record for this book is available from the British Library.

Trends in Abstract and Applied Analysis — Vol. 12
STIELTJES DIFFERENTIAL CALCULUS WITH APPLICATIONS

Copyright © 2025 by World Scientific Publishing Co. Pte. Ltd.

All rights reserved. This book, or parts thereof, may not be reproduced in any form or by any means, electronic or mechanical, including photocopying, recording or any information storage and retrieval system now known or to be invented, without written permission from the publisher.

For photocopying of material in this volume, please pay a copying fee through the Copyright Clearance Center, Inc., 222 Rosewood Drive, Danvers, MA 01923, USA. In this case permission to photocopy is not required from the publisher.

ISBN 978-981-12-9422-8 (hardcover)
ISBN 978-981-12-9423-5 (ebook for institutions)
ISBN 978-981-12-9424-2 (ebook for individuals)

For any available supplementary material, please visit
https://www.worldscientific.com/worldscibooks/10.1142/13869#t=suppl

Desk Editors: Nambirajan Karuppiah/Lai Fun Kwong

Typeset by Stallion Press
Email: enquiries@stallionpress.com

Preface

The notion of differentiating a function with respect to another function, called the Stieltjes differentiation, is a rather classical concept. It has roots at the genesis of calculus when, in an intuitive way, it was usual to think of a quantity varying continuously with respect to another (dependent or independent) quantity. The concept of derivatives of this type was already considered by Young (in 1917), Daniell (in 1918), and Ward (in 1936) in connection with the fundamental theorem of calculus for Stieltjes integrals. The Stieltjes derivative is a modification of the usual derivative through a nondecreasing left-continuous function. This change in the definition allows us to study several differential problems under the same framework. As a particular case, the Stieltjes derivative contains the Hilger delta derivative on time scales, thus offering a new unification and extension of the continuous and discrete calculus. Further, the study of differential equations in the sense of Stieltjes derivative allows the study of many classical problems in a unique framework. Also, this theory has the advantage that ordinary differential equations, ordinary difference equations, impulsive differential equations, dynamic equations on time scales, and generalized differential equations can be considered as particular instances of the Stieltjes differential equations. Therefore, it is important to study the Stieltjes differential calculus and qualitative properties of corresponding differential equations involving the Stieltjes derivative. Both the authors came across the topic of Stieltjes derivative while writing their book, *Generalized Quantum Calculus with Applications*. Inspired by the research works mentioned in the references section of this book, it was decided to come up with a book that covers the fundamentals of Stieltjes calculus and its applications, and will be useful for researchers to harness this powerful technique further to unlock new insights and embrace the intricacies of natural processes.

As the title suggests, this book is devoted to the recent developments in the study of Stieltjes derivative and its applications. This book provides an integrated approach to the study of some special aspects of Stieltjes derivatives and the qualitative theory of Stieltjes differential equations. The key topics include Stieltjes integral, Stieltjes derivative, elementary functions in the sense of Stieltjes, Laplace transform in the sense of Stieltjes, first-order and second-order linear Stieltjes differential equations, qualitative behaviour of systems of Stieltjes differential equations. The book is suitable for senior undergraduate students and beginning graduate students of engineering and science courses. The book consists of ten well-written research chapters reflecting the advances in the subject of Stieltjes differential calculus and Stieltjes differential equations. Each chapter contains a good percentage of new material. The chapters in the book are pedagogically organized. Each chapter concludes with a section with practical problems.

Chapter 1 deals with the Stieltjes integral and some of its properties. We first discuss the differentiation of monotone functions and then study the functions of finite variation and their properties along with the Helly principle of choice.

Chapter 2 is devoted to the concept of Stieltjes derivative. Basic definitions and examples are given. The main properties and main rules for the Stieltjes derivative are deduced. Moreover, the chain rules, mean value theorems, the Leibniz–Newton–Stieltjes formula, and the Stieltjes–Taylor formula are presented.

Chapter 3 assembles some Stieltjes elementary functions. Some elementary functions, such as Stieltjes exponential, Stieltjes trigonometric, and Stieltjes hyperbolic functions are defined and their basic properties are given.

In Chapter 4, we define Stieltjes–Laplace transform and prove some of its properties. Some special classes of functions which are useful for Stieltjes–Laplace transform are defined. Stieltjes–Laplace transforms for some Stieltjes elementary functions are presented.

Chapter 5 investigates the homogeneous and nonhomogeneous initial value problems for first-order Stieltjes differential equations. The integral representation of their solutions is given. The method of Stieltjes–Laplace transform for solving the initial value problems is presented.

In Chapter 6, we investigate homogeneous and nonhomogeneous second-order linear Stieltjes differential equations. The concept of Stieltjes Wronskian is introduced and some of their properties are given. The Stieltjes analogues of the Abel theorem is proved. A fundamental system of solutions is defined and using this, some representations of general solutions of the considered classes of Stieltjes differential equations are given. The concept of reducing of order and the method of factoring are discussed. The Stieltjes–Euler–Cauchy equations are investigated. The annihilator method and the Stieltjes–Laplace transform method are presented.

Chapter 7 deals with Stieltjes differential systems and their structures. The Stieltjes exponential matrix function is defined and some of its properties are deduced. The Stieltjes analogue of the classical Liouville theorem is proved. The Stieltjes Putzer algorithm for finding the Stieltjes exponential matrix function in the case when the matrix is constant is introduced. Stieltjes adjoint systems are introduced. The method of variation of constants is discussed for Stieltjes differential systems. The existence and uniqueness of solutions to nonlinear Stieltjes differential systems are also investigated.

In Chapter 8, the concept of periodicity for Stieltjes differential linear systems is introduced. The asymptotic behaviour of solutions of Stieltjes differential linear systems is discussed.

In Chapter 9, we investigate the stability for Stieltjes differential systems. We first provide the basic definitions and examples, and various criteria for the stability of solutions. The concept of uniform stability and sufficient conditions for it are then presented. The quasi-linear Stieltjes differential systems are defined, and their stability are investigated. The stability of two-dimensional autonomous systems is also presented.

Chapter 10 deals with the study of Stieltjes linear boundary value problems for both homogeneous and nonhomogeneous Stieltjes differential equations. The existence and uniqueness of solutions of these boundary value problems are investigated.

The aim of this book is to present a clear and well-organized treatment of the concept behind the development of mathematics as well as solution techniques involving the Stieltjes derivative. The contents of the book are presented in an easy-to-read and mathematically solid format.

Sanket Tikare would like to express special thanks to Dr. Usha Mukundan, Academic Director of Hindi Vidya Prachar Samiti, and Dr. Himashu Dawda, Principal of Ramniranjan Jhunjhunwala College, for their valuable and continued support, in a variety of ways, during the preparation of this book. Sanket also expresses his sincere gratitude to his co-author Svetlin G. Georgiev for the lively and productive discussions which led to an incredibly enriching experience, both professionally and personally. Further, Sanket extends his heartfelt appreciation to his wife Shaila and daughter Manasvini for their moral support and unflinching encouragement. Sanket has dedicated this book to his lifetime teacher Late Shri Manohar Hari Risbud.

The authors wish to acknowledge their families and friends for their constant support and encouragement. Finally, authors would like to thank the Editor-in-Chief of the monograph series: Trends in Abstract and Applied Analysis for the World Scientific Publishers, *John R. Graef*, for his interest and continued support to this project, and Executive Editor, *Rochelle Kronzek Miller* and desk editor *Lai Fun Kwong* for their genial attitude and co-operations at every stage in the preparation of the final version of this manuscript.

Svetlin G. Georgiev,
Paris, France
Sanket Tikare,
Mumbai, India

About the Authors

Svetlin G. Georgiev is a Professor of Mathematics in the Department of Mathematics at Sorbonne University, Paris, France. He works on various aspects of mathematics. Currently, he focuses on ordinary and partial differential equations, harmonic analysis, integral equations, Clifford and quaternion analysis, time scales calculus, and differential and dynamic geometry.

Sanket Tikare is an Assistant Professor in the Department of Mathematics at Ramniranjan Jhunjhunwala College, which is an empowered autonomous college affiliated with the University of Mumbai, Mumbai, Maharashtra, India. He did his MSc and PhD in mathematics from Shivaji University, Kolhapur, Maharashtra, India in the years 2008 and 2012, respectively. He specializes in applied analysis. His research areas of interest include time scale theory, differential equations, difference equations, quantum difference equations, dynamic equations on time scales, impulsive systems, and equations with delays.

Contents

Preface   v

About the Authors   ix

1. The Stieltjes Integral   1
    1.1 Monotone Functions 1
    1.2 Differentiation of Monotone Functions 6
    1.3 Functions of Bounded Variation 17
    1.4 The Helly Principle of Choice 29
    1.5 Continuous Functions of Bounded Variation 33
    1.6 The Stieltjes Integral 41
    1.7 Passage to the Limit Under the Stieltjes
          Integral Sign 52
    1.8 Advanced Practical Problems 56

2. The Stieltjes Derivative   59
    2.1 Definition and Examples 59
    2.2 Properties of the Stieltjes Derivative 64
    2.3 Chain Rules 70
    2.4 Mean Value Theorems 71
    2.5 Stieltjes Primitive and the Leibniz–
          Newton–Stieltjes Formula 79
    2.6 Stieltjes Monomials and the
          Stieltjes–Taylor Formula 81
    2.7 Advanced Practical Problems 89

3. Stieltjes Elementary Functions — 91

- 3.1 Stieltjes Regressive Functions 91
- 3.2 The Stieltjes Exponential Function 104
- 3.3 Stieltjes Trigonometric Functions 112
- 3.4 Stieltjes Hyperbolic Functions 114
- 3.5 Advanced Practical Problems 116

4. The Stieltjes–Laplace Transform — 119

- 4.1 Functions of Exponential Orders 119
- 4.2 Definition and Properties of the Stieltjes–Laplace Transform 121
- 4.3 The Stieltjes–Laplace Transform of Stieltjes Derivative 128
- 4.4 The Stieltjes–Laplace Transform of Stieltjes Integrals 132
- 4.5 Advanced Practical Problems 136

5. First-Order Linear Stieltjes Differential Equations — 139

- 5.1 Linear Stieltjes Differential Operator 139
- 5.2 Homogeneous Stieltjes Initial Value Problems 141
- 5.3 Nonhomogeneous Stieltjes Initial Value Problems .. 142
- 5.4 The Stieltjes–Laplace Transform Method 145
- 5.5 Advanced Practical Problems 148

6. Second-Order Linear Stieltjes Differential Equations — 151

- 6.1 Stieltjes–Wronskians 151
- 6.2 Homogeneous Second-Order Linear Stieltjes Differential Equations with Constant Coefficients .. 157
- 6.3 Reduction of Order 160
- 6.4 The Method of Factoring 161
- 6.5 Stieltjes–Euler–Cauchy Equations 163
- 6.6 Variation of Parameters 169
- 6.7 The Annihilator Method 173
- 6.8 The Stieltjes–Laplace Transform Method 175
- 6.9 Advanced Practical Problems 177

7.	**Stieltjes Differential Systems**	**181**
	7.1 Structure of Stieltjes Differential Systems	181
	7.2 The Stieltjes Matrix Exponential Function	202
	7.3 nth-Order Stieltjes Differential Equations	209
	7.4 Stieltjes Homogeneous Systems	215
	7.5 Stieltjes Fundamental Matrix Solutions	220
	7.6 Stieltjes Adjoint Differential Systems	227
	7.7 The Method of Variation of Constants	228
	7.8 Constant Coefficients	231
	7.9 Nonlinear Stieltjes Differential Systems	246
	7.10 Advanced Practical Problems	251
8.	**Qualitative Analysis of Stieltjes Differential Systems**	**255**
	8.1 Linear Periodic Stieltjes Differential Systems	255
	8.2 Asymptotic Behaviour of Solutions	265
	8.3 Advanced Practical Problems	266
9.	**Stability Theory for Stieltjes Differential Systems**	**271**
	9.1 Definition and Examples	271
	9.2 Criteria for Stability	273
	9.3 Uniform Stability	281
	9.4 Stability of Quasi-Linear Stieltjes Differential Systems	290
	9.5 Two-Dimensional Autonomous Stieltjes Differential Systems	297
	9.6 Advanced Practical Problems	330
10.	**Linear Stieltjes Boundary Value Problems**	**335**
	10.1 Introduction	335
	10.2 Existence of Solutions	341
	10.3 Advanced Practical Problems	345
	Bibliography	349
	Index	353

Chapter 1

The Stieltjes Integral

We discuss the concept of differentiation of monotone functions and the concept of bounded variations. The Helly principle of choice is presented. The Stieltjes integral which is essential for introducing the Stieltjes derivative are defined. We also discuss the process of limit under the Stieltjes integral sign.

Throughout this chapter, we assume f to be a real-valued function defined on $[a, b]$ and suppose that $[a, b] \subset \mathbb{R}$.

1.1 Monotone Functions

Definition 1.1. A function f defined on $[a, b]$ is said to be increasing (decreasing) provided

$$f(t) \leq f(s) \ (f(t) \geq f(s)) \quad \text{for } t < s, \quad t, s \in [a, b].$$

If

$$f(t) < f(s) \ (f(t) > f(s)) \quad \text{for } t < s, \quad t, s \in [a, b],$$

then f is said to be strictly increasing (strictly decreasing). Increasing functions and decreasing functions are called monotonic or monotone (strictly monotonic). If f decreases, then $-f$ increases. Thus, we can consider only increasing functions in many problems involving monotonic functions. Monotonic functions will always be considered finite.

Let f be an increasing function on $[a,b]$ and let $t_0 \in [a,b]$. Take $\{t_n\}_{n\in\mathbb{N}}$ so that $t_n > t_0$, $n \in \mathbb{N}$, and $\lim_{n\to\infty} t_n = t_0$. Then, $t_n > t_m$ for $n < m$, $n,m \in \mathbb{N}$, and

$$f(t_1) \geq f(t_n)$$
$$\geq f(t_m)$$
$$\geq f(t_0), \quad n < m, \quad n,m \in \mathbb{N}.$$

Thus, $\{f(t_n)\}_{n\in\mathbb{N}}$ is a decreasing bounded sequence. Then, $\lim_{n\to\infty} f(t_n)$ exists and it is finite. This limit is nothing but $\inf_{t\in(t_0,b]} f(t)$ and it does not depend on the choice of the sequence $\{t_n\}_{n\in\mathbb{N}}$. It is denoted by $f(t_0+)$. The symbol $f(t_0-)$ is defined similarly. We have that

$$f(t_0-) \leq f(t_0) \leq f(t_0+)$$

and

$$f(a) \leq f(a+),$$
$$f(b-) \leq f(b).$$

From here, we conclude that f is continuous at t_0 if and only if

$$f(t_0-) = f(t_0) = f(t_0+).$$

Definition 1.2. The numbers

$$f(t_0) - f(t_0-) \quad \text{and} \quad f(t_0+) - f(t_0)$$

are said to be the left and the right saltus, respectively, of f at t_0. The number

$$f(t_0+) - f(t_0-)$$

is said to be the saltus of f at t_0.

Theorem 1.1. *Let f be an increasing function on $[a,b]$ and $t_j \in (a,b)$, $j \in \{1,\ldots,n\}$. Then,*

$$(f(a+) - f(a)) + \sum_{j=1}^{n}(f(t_j+) - f(t_j-)) + (f(b) - f(b-))$$

$$\leq f(b) - f(a). \qquad (1.1)$$

Proof. Without any restriction, assume that

$$a = t_0 < t_1 < \cdots < t_n < t_{n+1} = b.$$

Note that, for any $j \in \{1,\ldots,n\}$, there is an s_j such that $t_j < s_j < t_{j+1}$. Then,
$$f(t_j+) \leq f(s_j),$$
$$f(t_j-) \geq f(s_{j-1}), \quad j \in \{1,\ldots,n\}.$$
Hence,
$$f(t_j+) - f(t_j-) \leq f(s_j) - f(s_{j-1}), \quad j \in \{1,\ldots,n\},$$
and
$$\sum_{j=1}^{n}(f(t_j+) - f(t_j-)) \leq \sum_{j=1}^{n}(f(s_j) - f(s_{j-1})), \quad j \in \{1,\ldots,n\}. \tag{1.2}$$

Next,
$$f(a+) \leq f(s_1), \quad f(b-) \geq f(s_n).$$
Then,
$$f(a+) - f(a) \leq f(s_0) - f(a),$$
$$f(b) - f(b-) \leq f(b) - f(s_n).$$
Now, adding (1.2) and the last two inequalities, we obtain
$$(f(a+) - f(a)) + \sum_{j=1}^{n}(f(t_j+) - f(t_j-)) + f(b) - f(b-)$$
$$\leq f(s_0) - f(a) + \sum_{j=1}^{n}(f(s_j) - f(s_{j-1})) + f(b) - f(s_n)$$
$$\leq f(s_0) - f(a) + (f(s_1) - f(s_0)) + (f(s_2) - f(s_1))$$
$$+ \cdots + (f(s_n) - f(s_{n-1})) + f(b) - f(s_n)$$
$$= f(b) - f(a),$$
i.e.,
$$(f(a+) - f(a)) + \sum_{j=1}^{n}(f(t_j+) - f(t_j-)) + f(b) - f(b-) \leq f(b) - f(a).$$
This completes the proof. □

Corollary 1.1. *An increasing function f defined on $[a,b]$ can have only a finite number of points of discontinuity at which its saltus is greater than a given positive number σ.*

Proof. Suppose f has n number of points of discontinuity at which its saltus is greater than a given positive number σ. Let $t_j \in [a,b]$, $j \in \{1,\ldots,n\}$, be such that
$$f(t_j+) - f(t_j-) \geq \sigma, \quad j \in \{1,\ldots,n\}.$$
Since f is increasing on $[a,b]$, we have that
$$f(b) - f(b-) \geq 0 \quad \text{and} \quad f(a+) - f(a) \geq 0.$$
Hence, applying (1.1), we arrive at
$$f(b) - f(a) \geq f(a+) - f(a) + \sum_{j=1}^{n}(f(t_j+) - f(t_j-)) + (f(b) - f(b-))$$
$$\geq \sum_{j=1}^{n}(f(t_j+) - f(t_j-))$$
$$\geq n\sigma.$$
Since $f(b) - f(a)$ is a finite number, in view of the last inequality, it follows that n must be finite. This completes the proof. □

Theorem 1.2. *The set of points of discontinuity of an increasing function f defined on $[a,b]$ is at most denumerable. Moreover, if $t_1, t_2, \ldots$ are all of the interior points of discontinuity, then*
$$(f(a+) - f(a)) + \sum_{j=1}^{\infty}(f(t_j+) - f(t_j-)) + (f(b) - f(b-)) \leq f(b) - f(a).$$
(1.3)

Proof. Let H be the set of all points of discontinuity for the function f. For $k \in \mathbb{N}$, let H_k denote the set of all points of discontinuity of f such that its saltus is greater than $1/k$. By Corollary 1.1, it follows that H_k is finite for $k \in \mathbb{N}$. Hence,
$$H = H_1 + H_2 + \cdots,$$
and H is at most denumerable. Next, letting $n \to \infty$ into the inequality (1.1), we get the inequality (1.3). This completes the proof. □

Definition 1.3. Let f be an increasing function on $[a,b]$. Define the saltus function $S^f : [a,b] \to \mathbb{R}$ as

$$S^f(t) = \begin{cases} 0, & t = a, \\ (f(a+) - f(a)) + \sum_{t_j < t} (f(t_j+) & t \in (a,b]. \\ \quad - f(t_j-)) + (f(t) - f(t-)), & \end{cases}$$

Theorem 1.3. *Let f be an increasing function defined on $[a,b]$. Then the function $\phi = f - S^f$ is continuous and increasing function on $[a,b]$.*

Proof. Let $a \leq t < s \leq b$. Applying (1.3) for $[t,s]$ instead of the interval $[a,b]$, we get

$$S^f(s) - S^f(t) \leq f(s) - f(t), \tag{1.4}$$

whereupon

$$f(t) - S^f(t) \leq f(s) - S^f(s),$$

i.e.,

$$\phi(t) \leq \phi(s).$$

Thus, ϕ is an increasing function on $[a,b]$. Now, let s approach t into (1.4). Then,

$$S^f(t+) - S^f(t) \leq f(t+) - f(t). \tag{1.5}$$

On the other hand, by the definition of S^f, for $t < s$, we find

$$S^f(s) = (f(a+) - f(a)) + \sum_{t_j < s} (f(t_j+) - f(t_j-)) + (f(s) - f(s-))$$

$$\geq (f(a+) - f(a)) + \sum_{t_j < t} (f(t_j+) - f(t_j-)) + (f(t) - f(t-))$$

$$+ (f(t+) - f(t))$$

$$= S^f(t) + f(t+) - f(t),$$

whereupon

$$f(t+) - f(t) \leq S^f(s) - S^f(t),$$

and then,
$$f(t+) - f(t) \leq S^f(t+) - S^f(t).$$
Hence, using (1.5), we find
$$f(t+) - f(t) = S^f(t+) - S^f(t),$$
and therefore,
$$\phi(t+) = \phi(t).$$
In a similar way, we obtain
$$\phi(t-) = \phi(t).$$
Thus, ϕ is a continuous function on $[a, b]$. This completes the proof. $\square$

1.2 Differentiation of Monotone Functions

Let A be a given abstract set and let f be a given function defined on A. Then, we denote
$$f(E) = \{f(t) : t \in E\}.$$
Thus, the function f induces a mapping of the family of all subsets of A into the family of all subsets of the image set $f(A)$. In other words, $f(E)$ consists of those elements s for which the equation $f(t) = s$ has at least one solution in the set E.

Theorem 1.4. *If $E_1 \subset E_2 \subset A$, then*
$$f(E_1) \subset f(E_2). \tag{1.6}$$

Proof. Let $s \in f(E_1)$ be arbitrarily chosen. Then there is an element $t \in E_1$ so that $s = f(t)$. Since $E_1 \subset E_2$, we have that $t \in E_2$. Thus, $s \in f(E_2)$. Since $s \in f(E_1)$ was arbitrarily chosen and we get that it is an element of $f(E_2)$, we arrive at the inclusion (1.6). This completes the proof. $\square$

Theorem 1.5. *Let $E_j \subset A$, $j \in \mathbb{N}$. Then,*
$$f(\cup_{j=1}^{\infty} E_j) = \cup_{j=1}^{\infty} f(E_j). \tag{1.7}$$

Proof. Let $s \in f(\cup_{j=1}^{\infty} E_j)$ be arbitrarily chosen. Then, there is $t \in \cup_{j=1}^{\infty} E_j$ so that $s = f(t)$. Since $t \in \cup_{j=1}^{\infty} E_j$, there exists an $l \in \mathbb{N}$ such that $t \in E_l$. Hence, $s \in f(E_l)$ and $s \in \cup_{j=1}^{\infty} f(E_j)$. Because $s \in f(\cup_{j=1}^{\infty} E_j)$ was arbitrarily chosen and we get that it is an element of $\cup_{j=1}^{\infty} f(E_j)$, we arrive at the inclusion

$$f\left(\cup_{j=1}^{\infty} E_j\right) \subset \cup_{j=1}^{\infty} f(E_j). \tag{1.8}$$

Now, let $u \in \cup_{j=1}^{\infty} f(E_j)$ be arbitrarily chosen. Then there is an $m \in \mathbb{N}$ so that $u \in f(E_m)$. Hence there exists a $v \in E_m$ for which $u = f(v)$ and $v \in \cup_{j=1}^{\infty} E_j$. Therefore, $u \in f(\cup_{j=1}^{\infty} E_j)$. Since $u \in \cup_{j=1}^{\infty} f(E_j)$ was arbitrarily chosen and we get that it is an element of $f(\cup_{j=1}^{\infty} E_j)$, we find the inclusion

$$\cup_{j=1}^{\infty} f(E_j) \subset f\left(\cup_{j=1}^{\infty} E_j\right).$$

From this inclusion and the inclusion (1.8), we find (1.7). This completes the proof. □

Example 1.1. Let $A = [a, b]$ and $f \colon [a, b] \to \mathbb{R}$ be a continuous and increasing function. Then $f(A) = [f(a), f(b)]$.

Exercise 1.1. Let $E_j \subset A$, $j \in \mathbb{N}$, and f be defined on A. Then prove that

$$f(\cap_{j=1}^{\infty} E_j) = \cap_{j=1}^{\infty} f(E_j).$$

Exercise 1.2. Let $E_1, E_2 \subset A$ with $E_1 \cap E_2 = \emptyset$ and f be defined on A. Then prove that $f(E_1) \cap f(E_2) = \emptyset$.

Definition 1.4. The number λ is said to be a derived number of f at t_0 provided there exists a sequence $\{h_n\}_{n \in \mathbb{N}}$, $h_n \neq 0$, $n \in \mathbb{N}$, such that $h_n \to 0$ as $n \to \infty$, and

$$\lambda = \lim_{n \to \infty} \frac{f(t_0 + h_n) - f(t_0)}{h_n}.$$

Obviously, the derived number is denoted by $Df(t_0)$.

Example 1.2. Let $f'(t_0)$ exists. Then $Df(t_0) = f'(t_0)$.

Example 1.3. Consider the Dirichlet function

$$f(t) = \begin{cases} 1 & \text{for } t \in \mathbb{Q}, \\ 0 & \text{for } t \in \mathbb{R} \setminus \mathbb{Q}. \end{cases}$$

Let $h, t_0 \in \mathbb{R}$ be arbitrarily chosen. We have the following cases:

(1) Let $t_0 \in \mathbb{Q}$. Then,

$$\frac{f(t_0 + h) - f(t_0)}{h} = \begin{cases} 0 & \text{for } h \in \mathbb{Q}, \\ -\frac{1}{h} & \text{for } h \in \mathbb{R} \setminus \mathbb{Q}. \end{cases}$$

(2) Let $x_0 \in \mathbb{R} \setminus \mathbb{Q}$. Then,

$$\frac{f(t_0 + h) - f(t_0)}{h} = \begin{cases} 0 & \text{for } t_0 + h, \ h \in \mathbb{R} \setminus \mathbb{Q}, \\ \frac{1}{h} & \text{for } t_0 + h \in \mathbb{Q}. \end{cases}$$

Thus, the Dirichlet function has three derived numbers: $0, \infty$, and $-\infty$.

Theorem 1.6. *Let the function f be defined on $[a, b]$. Then, its derived numbers exist at every point in $[a, b]$.*

Proof. Let $t_0 \in [a, b]$ and $\{h_n\}_{n \in \mathbb{N}}$, $h_n \neq 0$, be a sequence such that $h_n \to 0$ as $n \to \infty$, and $t_0 + h_n \in [a, b]$. Set

$$\sigma_n = \frac{f(t_0 + h_n) - f(t_0)}{h_n}, \quad n \in \mathbb{N}.$$

If sequence $\{\sigma_n\}_{n \in \mathbb{N}}$ is bounded, then, applying the Bolzano–Weierstrass theorem, we conclude that there is a subsequence $\{\sigma_{n_k}\}_{k \in \mathbb{N}}$ that is convergent to a derived number of f at x_0. If sequence $\{\sigma_n\}_{n \in \mathbb{N}}$ is unbounded, then there is a subsequence $\{\sigma_{n_k}\}_{k \in \mathbb{N}}$ so that $\sigma_{n_k} \to \pm\infty$ as $k \to \infty$, and in this case, $Df(t_0) = \pm\infty$. This completes the proof. $\square$

Theorem 1.7. *A function f defined on $[a, b]$ has derivative $f'(t_0)$ at $t_0 \in [a, b]$ if and only if all derived numbers of f at t_0 are equal.*

Proof. Suppose all derived numbers of f at t_0 are equal to λ. We will prove that for any sequence $\{h_n\}_{n \in \mathbb{N}}$, $h_n \neq 0$, we have $h_n \to 0$ as $n \to \infty$, and

$$\lim_{n \to \infty} \frac{f(t_0 + h_n) - f(t_0)}{h_n} = \lambda.$$

Assume that this is not the case. Then there is a sequence $\{h_n\}_{n\in\mathbb{N}}$, $h_n \neq 0$, so that the sequence

$$\sigma_n = \frac{f(t_0 + h_n) - f(t_0)}{h_n}, \quad n \in \mathbb{N},$$

does not have the limit λ. Let $\lambda \in (-\infty, \infty)$. Then there is an $\varepsilon > 0$ such that there is an infinite set of numbers σ_n that lie outside the interval $(\lambda - \varepsilon, \lambda + \varepsilon)$. This infinite set contains a subsequence $\{\sigma_{n_k}\}_{k\in\mathbb{N}}$ which tends to a finite or infinite limit μ and $\mu \neq \lambda$. This is a contradiction. Hence all derived numbers of f at t_0 are equal to λ. The case $\lambda = \pm\infty$ is evident and we leave it to the reader as an exercise.

Conversely, suppose $f'(t_0)$ exists. This part we leave it to the reader as an exercise. □

Theorem 1.8. *Let f be a monotonically increasing function defined on $[a, b]$. If, at every point t of $E \subset [a, b]$, there exists at least one derived number $Df(t)$ such that*

$$Df(t) \leq p$$

for some $p > 0$, then

$$m^*(f(E)) \leq p m^*(E). \tag{1.9}$$

Here $m^*(A)$, $m_*(A)$, and $m(A)$ denote the outer measure of A, inner measure of A, and the measure of A, respectively.

Proof. Let $\varepsilon > 0$ be arbitrarily chosen. We choose a bounded open set G so that

$$E \subset G \quad \text{and} \quad m(G) < m^*(E) + \varepsilon.$$

Also, let $p_0 > p$. If $t_0 \in E$, then there exists a sequence $\{h_n\}_{n\in\mathbb{N}}$, $h_n \neq 0$, $h_n \to 0$ as $n \to \infty$, such that

$$\lim_{n\to\infty} \frac{f(t_0 + h_n) - f(t_0)}{h_n} = Df(t_0) \leq p.$$

Take $N \in \mathbb{N}$ sufficiently large so that $[t_0, t_0 + h_n] \subset G$, $n > N$, and

$$\frac{f(t_0 + h_n) - f(t_0)}{h_n} < p_0, \quad n > N.$$

Set

$$d_n(t_0) = [t_0, t_0 + h_n],$$
$$\Delta_n(t_0) = [f(t_0), f(t_0 + h_n)], \quad n \in \mathbb{N}.$$

Since f is increasing, we have
$$f(d_n(t_0)) \subset \Delta_n(t_0), \quad n > N.$$
Next,
$$m(d_n(t_0)) = |h_n|,$$
$$m(\Delta_n(t_0)) = |f(t_0 + h_n) - f(t_0)|, \quad n > N,$$
and
$$|f(t_0 + h_n) - f(t_0)| < p_0|h_n|, \quad n > N.$$
Thus,
$$m(\Delta_n(t_0)) < p_0 m(d_n(t_0)), \quad n > N.$$
Since $h_n \to 0$ as $n \to \infty$, by the last inequality, it follows that among the intervals $\Delta_n(t_0)$, there is an arbitrary small one. Note that $f(E)$ is covered in the sense of Vitali by the intervals $\Delta_n(t)$, $n \in \mathbb{N}$, $t \in E$. Now, applying the Vitali theorem, we can select from the family
$$\{\Delta_n(t) : t \in E, \, n \in \mathbb{N}\},$$
a countable sequence $\{\Delta_{n_l}(t_l)\}_{l \in \mathbb{N}}$ of pairwise disjoint intervals such that $\{d_{n_l}(t_l)\}_{l \in \mathbb{N}}$ are pairwise disjoint and
$$m\left(f(E) - \sum_{l=1}^{\infty} \Delta_{n_l}(t_l)\right) = 0.$$
Hence,
$$m^*(f(E)) \leq \sum_{l=1}^{\infty} m(\Delta_{n_l}(t_l))$$
$$< p_0 \sum_{l=1}^{\infty} m(d_{n_l}(t_l)) \qquad (1.10)$$
$$= p_0 m\left(\sum_{l=1}^{\infty} d_{n_l}(t_l)\right).$$
Because
$$\sum_{l=1}^{\infty} d_{n_l}(t_l) \subset G,$$

we find
$$m\left(\sum_{l=1}^{\infty} d_{n_l}(t_l)\right) \leq m(G)$$
$$< m^*(E) + \varepsilon.$$
Hence, applying (1.10), we find
$$m^*(f(E)) \leq p_0(m^*(E) + \varepsilon).$$
Letting $\varepsilon \to 0$, we get (1.9). This completes the proof. □

Theorem 1.9. *Let f be a strictly increasing function defined on $[a,b]$. If at every point t of $E \subset [a,b]$, there exists at least one derived number, $Df(t)$ such that*
$$Df(t) \geq q$$
for some $q \geq 0$, then
$$m^*(f(E)) \geq qm^*(E). \qquad (1.11)$$

Proof. For $q = 0$, we always have
$$m^*(f(E)) \geq 0.$$
Therefore, suppose that $q > 0$. Take $q_0 > 0$ so that $q > q_0$. Let $\varepsilon > 0$ be arbitrarily chosen. Also, let G be a bounded open set so that
$$f(E) \subset G \quad \text{and} \quad m(G) < m^*(f(E)) + \varepsilon.$$
Denote with E_c the set of all points of E at which f is continuous. By Theorem 1.2, it follows that $E \setminus E_c$ is at most denumerable. Let $t_0 \in E$. Then, there is a sequence $\{h_n\}_{n \in \mathbb{N}}$, $h_n \neq 0$, and $h_n \to 0$ as $n \to \infty$, such that
$$\lim_{n \to \infty} \frac{f(t_0 + h_n) - f(t_0)}{h_n} = Df(t_0) \geq q.$$
Without any restriction, assume that
$$\frac{f(t_0 + h_n) - f(t_0)}{h_n} > q_0, \quad n \in \mathbb{N}.$$
Let $d_n(t_0)$ and $\Delta_n(t_0)$ be as in the proof of Theorem 1.8. Then we have
$$m(\Delta_n(t_0)) > q_0 m(d_n(t_0)).$$
If $t_0 \in E_c$, then the whole interval $[f(t_0), f(t_0 + h_n)]$ lie entirely in the set G for sufficiently large n. Without any restriction, assume

that this is the case for all $n \in \mathbb{N}$. Then E_c is covered in the sense of Vitali by the intervals $d_n(t_0)$, $n \in \mathbb{N}$. By the Vitali theorem, it follows that there exists a denumerable sequence of pairwise disjoint intervals $\{d_{n_j}(t_j)\}_{j \in \mathbb{N}}$ such that $\{\Delta_{n_j}(t_0)\}_{j \in \mathbb{N}}$ are pairwise disjoint and

$$m\left(E_c - \sum_{j=1}^{\infty} d_{n_j}(t_j)\right) = 0.$$

Hence,

$$m^*(E_c) \leq \sum_{j=1}^{\infty} m(d_{n_j}(t_j))$$

$$< \frac{1}{q_0} \sum_{j=1}^{\infty} m(\Delta_{n_j}(t_j))$$

$$= \frac{1}{q_0} m\left(\sum_{j=1}^{\infty} \Delta_{n_j}(t_j)\right)$$

$$= \frac{1}{q_0} m(G)$$

$$\leq \frac{1}{q_0}\left(m^*(f(E)) + \varepsilon\right).$$

Letting $\varepsilon \to 0$, we get (1.11). This completes the proof. □

Corollary 1.2. *The set of points of which at least one derived number of an increasing function defined on $[a, b]$ is infinite is of measure zero.*

Proof. Let f be strictly increasing and $E(Df(t) = \infty)$ denotes the set of points x in E at which at least one derived number is ∞. Then

$$f(E(Df(t) = \infty)) \subset [f(a), f(b)]$$

and

$$m^*(f(E(Df(t) = \infty))) < \infty.$$

Assume that

$$m^*(E(Df(t) = \infty)) > 0.$$

Then, applying Theorem 1.9, we get that
$$m^*(f(E(Df(t) = \infty))) = \infty.$$
This is a contradiction. Next, suppose f is increasing. Then, we define
$$F(t) = f(t) + t, \quad x \in [a, b],$$
which is strictly increasing on $[a, b]$. Note that for any $h \in \mathbb{R}$, $h \neq 0$, we have
$$\frac{F(t + h) - F(t)}{h} = \frac{f(t + h) + t + h - f(t) - t}{h}$$
$$= \frac{f(t + h) - f(t)}{h} + 1, \quad t \in [a, b].$$
Thus, the set of points $t \in [a, b]$ at which $Df(t) = \infty$ coincides with the same set of F, i.e., $m^*(E(DF(t) = \infty)) = m^*(E(Df(t) = \infty))$. Since F is strictly increasing, we have $m^*(E(DF(t) = \infty)) = 0$. Hence,
$$m^*(E(Df(t) = \infty)) = 0.$$
This completes the proof. □

Theorem 1.10. *Let f be an increasing function defined on $[a, b]$ and let p and q be two numbers such that $p < q$ and $E_{p,q} = \{t \in [a, b]: p < t < q\}$. If at every point t of the set $E_{p,q} \subset [a, b]$, there exist two derived numbers $D_1 f(t)$ and $D_2 f(t)$ such that*
$$D_1 f(t) < p < q < D_2 f(t),$$
then $m(E_{p,q}) = 0$.

Proof. Let f be a strictly increasing function. Then, by Theorem 1.8, we find
$$m^*(f(E_{p,q})) \leq p m^*(E_{p,q}),$$
and by Theorem 1.9, it follows that
$$m^*(f(E_{p,q})) \geq q m^*(E_{p,q}).$$
Thus,
$$q m^*(E_{p,q}) \leq p m^*(E_{p,q}).$$

Since $p < q$, we find
$$m^*(E_{p,q}) = 0. \tag{1.12}$$
Let f be an increasing function. Then, we consider the function F defined in the proof of Corollary 1.2 and we get
$$(q+1)m^*(E_{p,q}) \leq (p+1)m^*(E_{p,q}),$$
whereupon we find (1.12). This completes the proof. $\square$

Theorem 1.11. *An increasing function f defined on the interval $[a, b]$ has a finite derivative at almost all points of $[a, b]$.*

Proof. Let E be the set of those points of $[a, b]$ at which the derivative of f does not exist. Take $t_0 \in E$ arbitrarily. Then, there are two distinct derived numbers $D_1 f(t_0)$ and $D_2 f(t_0)$. Suppose that
$$D_1 f(t_0) < D_2 f(t_0).$$
Then there are two rational numbers p and q so that
$$D_1 f(t_0) < p < q < D_2 f(t_0).$$
Let $E_{p,q}$ be the sets in Theorem 1.10. Then,
$$E = \bigcup_{(p,q)} E_{p,q}.$$
By Theorem 1.10, we get
$$m(E_{p,q}) = 0$$
and then $m^*(E) = 0$. By Corollary 1.2, it follows that the set of points of $[a, b]$ at which at least one derived number is infinite is of measure zero. This completes the proof. $\square$

Theorem 1.12. *Let f be an increasing function defined on $[a, b]$. Then its derivative f' is measurable and*
$$\int_a^b f'(t)dt \leq f(b) - f(a),$$
i.e., f' is summable.

Proof. We extend the definition of f as follows:
$$f(t) = f(b), \quad t \in (b, b+1].$$

Then, at every point $t \in [a, b]$, we have

$$f'(t) = \lim_{n\to\infty} n\left(f\left(t+\frac{1}{n}\right) - f(t)\right).$$

That is, for $t \in [a, b]$, $f'(t)$ is the limit of an almost everywhere convergent sequence of measurable functions. Hence, f' is measurable. Since $f' \geq 0$ on $[a, b]$, we can consider its Lebesgue integral

$$\int_a^b f'(t)dt.$$

By the Fatou theorem, it follows that

$$\int_a^b f'(t)dt \leq \sup\left(n\int_a^b \left(f\left(t+\frac{1}{n}\right) - f(t)\right)dt\right)$$

$$= \sup\left(n\left(\int_a^b f\left(t+\frac{1}{n}\right)dt - \int_a^b f(t)dt\right)\right)$$

$$= \sup\left(n\left(\int_{a+\frac{1}{n}}^{b+\frac{1}{n}} f(s)ds - \int_a^b f(s)ds\right)\right)$$

$$= \sup\left(n\left(\int_{a+\frac{1}{n}}^a f(s)ds + \int_a^b f(s)ds + \int_b^{b+\frac{1}{n}} f(s)ds\right.\right.$$

$$\left.\left. - \int_a^b f(s)ds\right)\right)$$

$$= \sup\left(n\left(\int_b^{b+\frac{1}{n}} f(t)dt - \int_a^{a+\frac{1}{n}} f(t)dt\right)\right)$$

$$= \sup\left(n\left(\frac{1}{n}f(b) - \int_a^{a+\frac{1}{n}} f(t)dt\right)\right)$$

$$\leq \sup\left(n\left(\frac{1}{n}f(b) - \frac{1}{n}f(a)\right)\right)$$

$$= f(b) - f(a).$$

This completes the proof. □

Theorem 1.13. *For every subset E of measure zero on the closed interval $[a,b]$, there exists a continuous increasing function σ such that $\sigma'(t) = \infty$, $t \in E$.*

Proof. For $n \in \mathbb{N}$, let G_n denote a bounded open set such that
$$G_n \subset E, \quad m(G_n) < \frac{1}{2^n}.$$
Define
$$\psi_n(t) = m(G_n \cap [a,t]), \quad n \in \mathbb{N}, \quad t \in [a,b].$$
Then, ψ_n is an increasing function that is nonnegative, continuous, and satisfies the inequality
$$\psi_n(t) < \frac{1}{2^n}, \quad t \in [a,b].$$
Set
$$\sigma(t) = \sum_{n=1}^{\infty} \psi_n(t), \quad t \in E.$$
Then, σ is increasing, nonnegative and continuous. Fix $n \in \mathbb{N}$. Take $t_0 \in E$ arbitrarily. Let $|h|$ be sufficiently small so that $[t_0, t_0+h] \subset G_n$. Without any restriction, suppose that $h > 0$. Then,
$$\psi_n(t_0 + h) = m\left(G_n \cap [a, t_0 + h]\right)$$
$$= m\left((G_n \cap [a, t_0]) \cup (G_n \cap [t_0, x_0 + h])\right)$$
$$= m\left(G_n \cap [a, t_0]\right) + m\left(G_n \cap [t_0, t_0 + h]\right)$$
$$= \psi_n(t_0) + m([t_0, t_0 + h])$$
$$= \psi_n(t_0) + h,$$
whereupon
$$\frac{\psi(t_0 + h) - \psi(t_0)}{h} = 1.$$
Thus, for sufficiently small h and $N \in \mathbb{N}$, we have
$$\frac{\sigma(t_0 + h) - \sigma(t_0)}{h} \geq \sum_{n=1}^{N} \frac{\psi(t_0 + h) - \psi(t_0)}{h}$$
$$= N,$$
and then, $\sigma'(t_0) = \infty$. This completes the proof. $\square$

1.3 Functions of Bounded Variation

Suppose that f is a function defined on the bounded closed interval $[a, b]$ and P is a partition of $[a, b]$ which subdivide $[a, b]$ into parts by means of the points

$$P\colon a = t_0 < t_1 < \cdots < t_n = b, \quad n \in \mathbb{N}. \tag{1.13}$$

Consider the sum

$$\bigvee(f, P) = \sum_{j=0}^{n-1} |f(t_{j+1}) - f(t_j)|.$$

Definition 1.5. The least upper bound of the set of all possible sums $\bigvee(f, P)$ is called the total variation of f on $[a, b]$. It is denoted by $\bigvee_a^b(f)$. The function f is said to be a function of bounded variation on $[a, b]$ provided $\bigvee_a^b(f) < \infty$. We also state that f has finite variation on $[a, b]$.

Note 1.1. A function of bounded is also known as a function of finite variation.

Remark 1.1. A function of bounded variation need not be continuous. For example, $f(t) = \lfloor t \rfloor$, where $\lfloor t \rfloor$ denotes the greatest integer not greater than x, is a function of bounded variation on $[0, 2]$ but not continuous. Also, a continuous function may not be of bounded variation, see example in Remark 1.2.

Theorem 1.14. *A monotonic function f on $[a, b]$ is of bounded variation on $[a, b]$.*

Proof. Without any restriction, suppose that f is increasing on $[a, b]$. Then, for any partition (1.13) of $[a, b]$, we have

$$\bigvee(f, P) = \sum_{j=0}^{n-1} |f(t_{j+1}) - f(t_j)|$$

$$= \sum_{j=0}^{\infty} (f(t_{j+1}) - f(t_j))$$

$$= f(t_1) - f(a) + f(t_2) - f(t_1) + \cdots + f(b) - f(t_{n-1})$$

$$= f(b) - f(a)$$
$$< \infty.$$

This completes the proof. □

Definition 1.6. A finite function f defined on $[a,b]$ is said to satisfy the Lipschitz condition with constant $K > 0$ provided
$$|f(t) - f(s)| \leq K|t-s| \quad \text{for} \quad t, s \in [a,b].$$

Theorem 1.15. *Let the function f be defined on $[a,b]$ and satisfy the Lipschitz condition with constant $K > 0$ on $[a,b]$. Then f is of bounded variation on $[a,b]$.*

Proof. For any partition (1.13) of $[a,b]$, we have
$$\bigvee(f, P) = \sum_{j=0}^{\infty} |f(t_{j+1}) - f(t_j)|$$
$$\leq K \sum_{j=0}^{\infty} (t_{j+1} - t_j)$$
$$= K(t_1 - a + t_2 - t_1 + \cdots + b - t_{n-1})$$
$$= K(b - a)$$
$$< \infty.$$

This completes the proof. □

Theorem 1.16. *A function f of bounded variation on $[a,b]$ is bounded on $[a,b]$.*

Proof. Since f is a function of bounded variation on $[a,b]$, we have $\bigvee_a^b(f) < \infty$. Then,
$$|f(t) - f(a)| + |f(b) - f(t)| \leq \bigvee_a^b(f), \quad t \in [a,b],$$

whereupon
$$|f(t)| - |f(a)| \leq |f(t) - f(a)|$$
$$\leq \bigvee_a^b(f), \quad t \in [a,b],$$
$$|f(t)| \leq |f(a)| + \bigvee_a^b(f), \quad t \in [a,b].$$

This completes the proof. □

Remark 1.2. A bounded function need not be of bounded variation. For example,
$$f(t) = \begin{cases} t \sin\left(\frac{1}{t}\right) & \text{for } 0 < t \leq 1, \\ 0 & \text{for } t = 0. \end{cases}$$

Then f is bounded on $[0,1]$ but it is not of bounded variation on $[0,1]$. To see this, fix $n \in \mathbb{N}$ and let
$$P: 0 = t_0 < t_1 < \cdots < t_n = 1.$$
Note that $f(t_k) = (-1)^k t_k$, and so
$$|f(t_{k+1}) - f(t_k)| = t_{k+1} - t_k$$
$$\geq 2t_{k+1}$$
$$\geq \frac{4}{3\pi} \frac{1}{k+1}.$$

Consequently,
$$\bigvee(f,P) \geq \frac{4}{3\pi} \sum_{k=0}^{n-1} \frac{1}{k+1} \to \infty \quad \text{as } n \to \infty.$$

Theorem 1.17. *The sum, difference, and multiplication of two functions f and g of bounded variation on $[a,b]$ are functions of bounded variation on $[a,b]$.*

Proof. Since f and g are functions of bounded variation on $[a,b]$,

$$\bigvee_a^b(f) < \infty \quad \text{and} \quad \bigvee_a^b(g) < \infty.$$

Also, by Theorem 1.16, it follows that there exists a positive constant A such that

$$|f(t)| \leq A \quad \text{and} \quad |g(t)| \leq A, \quad t \in [a,b].$$

Then,

$$\bigvee(f+g, P) = \sum_{j=0}^{\infty} |(f+g)(t_{j+1}) - (f+g)(t_j)|$$

$$= \sum_{j=0}^{\infty} |f(t_{j+1}) + g(t_{j+1}) - (f(t_j) + g(t_j))|$$

$$= \sum_{j=0}^{\infty} |(f(t_{j+1}) - f(t_j)) + (g(t_{j+1}) - g(t_j))|$$

$$\leq \sum_{j=0}^{\infty} |f(t_{j+1}) - f(t_j)| + \sum_{j=0}^{\infty} |g(t_{j+1}) - g(t_j)|$$

$$= \bigvee(f, P) + \bigvee(g, P)$$

$$\leq \bigvee_a^b(f) + \bigvee_a^b(g),$$

whereupon

$$\bigvee_a^b(f+g) < \infty.$$

Next,

$$\bigvee(f-g, P) = \sum_{j=0}^{\infty} |(f-g)(t_{j+1}) - (f-g)(t_j)|$$

$$= \sum_{j=0}^{\infty} |f(t_{j+1}) - g(t_{j+1}) - (f(t_j) - g(t_j))|$$

$$= \sum_{j=0}^{\infty} |(f(t_{j+1}) - f(t_j)) - (g(t_{j+1}) - g(t_j))|$$

$$\leq \sum_{j=0}^{\infty} |f(t_{j+1}) - f(t_j)| + \sum_{j=0}^{\infty} |g(t_{j+1}) - g(t_j)|$$

$$= \bigvee(f, P) + \bigvee(g, P)$$

$$\leq \bigvee_a^b(f) + \bigvee_a^b(g)$$

and

$$\bigvee_a^b(f - g) < \infty.$$

Moreover,

$$\bigvee(fg, P) = \sum_{j=0}^{\infty} |(fg)(t_{j+1}) - (fg)(t_j)|$$

$$= \sum_{j=0}^{\infty} |f(t_{j+1})g(t_{j+1}) - (f(t_j)g(t_j))|$$

$$= \sum_{j=0}^{\infty} |(f(t_{j+1}) - f(t_j))g(t_{j+1}) + (g(t_{j+1}) - g(t_j))f(t_j)|$$

$$\leq \sum_{j=0}^{\infty} |f(t_{j+1}) - f(t_j)||g(t_{j+1})|$$

$$+ \sum_{j=0}^{\infty} |g(t_{j+1}) - g(t_j)||f(t_j)|$$

$$= A \left(\bigvee(f, P) + \bigvee(g, P) \right)$$

$$\leq A \left(\bigvee_a^b(f) + \bigvee_a^b(g) \right),$$

whereupon
$$\bigvee_a^b (fg) < \infty.$$

This completes the proof. □

Theorem 1.18. *Let f and g be functions of bounded variation on $[a,b]$ and $g \geq \sigma > 0$ on $[a,b]$. Then $\frac{f}{g}$ is a function of bounded variation on $[a,b]$.*

Proof. Since f and g are functions of bounded variation on $[a,b]$, we have
$$\bigvee_a^b (f) < \infty \quad \text{and} \quad \bigvee_a^b (g) < \infty.$$

Also, by Theorem 1.16, it follows that there exists a positive constant A such that
$$|f(t)| \leq A, \quad t \in [a,b].$$

Then,
$$\bigvee \left(\frac{f}{g}, P\right) = \sum_{j=0}^{\infty} \left| \left(\frac{f}{g}\right)(t_{j+1}) - \left(\frac{f}{g}\right)(t_j) \right|$$

$$= \sum_{j=0}^{n-1} \left| \frac{f(t_{j+1})}{g(t_{j+1})} - \frac{f(t_j)}{g(t_j)} \right|$$

$$= \sum_{j=0}^{n-1} \left| \frac{f(t_{j+1})g(t_j) - f(t_j)g(t_{j+1})}{g(t_j)g(t_{j+1})} \right|$$

$$\leq \frac{1}{\sigma^2} \sum_{j=0}^{n-1} |(f(t_{j+1}) - f(t_j))g(t_j) - f(t_j)(g(t_{j+1}) - g(t_j))|$$

$$\leq \frac{1}{\sigma^2} \left(\sum_{j=0}^{\infty} |f(t_{j+1}) - f(t_j)||g(t_j)| \right.$$

$$\left. + \sum_{j=0}^{n-1} |f(t_j)||g(t_{j+1}) - g(t_j)| \right)$$

$$\leq \frac{A}{\sigma^2}\left(\sum_{j=0}^{n-1}|f(t_{j+1})-f(t_j)|+\sum_{j=0}^{n-1}|g(t_{j+1})-g(t_j)|\right)$$

$$=\frac{A}{\sigma^2}\left(\bigvee(f,P)+\bigvee(g,P)\right)$$

$$\leq \frac{A}{\sigma^2}\left(\bigvee_a^b(f)+\bigvee_a^b(g)\right).$$

Therefore,

$$\bigvee_a^b\left(\frac{f}{g}\right)<\infty.$$

This completes the proof. □

Theorem 1.19. *Let f be a function of bounded variation on $[a,b]$ and $c \in (a,b)$. Then,*

$$\bigvee_a^b(f) = \bigvee_a^c(f) + \bigvee_c^b(f). \tag{1.14}$$

Proof. Let

$$P_1: a=s_0<s_1<\cdots<s_m=c$$

and

$$P_2: c=u_0<u_1<\cdots<u_n=b$$

be partitions of $[a,c]$ and $[c,b]$, respectively. Then

$$\bigvee_a^b(f) \geq \bigvee(f, P_1 \cup P_2)$$

$$=\sum_{j=0}^{m-1}|f(s_{j+1})-f(s_j)|+\sum_{k=0}^{n-1}|f(u_{k+1})-f(u_k)|$$

$$=\bigvee(f,P_1)+\bigvee(f,P_2),$$

whereupon

$$\bigvee_a^c(f)+\bigvee_c^b(f) \leq \bigvee_a^b(f). \tag{1.15}$$

Next, consider the partition
$$P_3: a = t_0 < t_1 < \cdots < t_p = c < \cdots < t_q = b$$
of the interval $[a, b]$. Then,
$$\bigvee(f, P_3) = \sum_{j=0}^{p-1} |f(t_{j+1}) - f(t_j)| + \sum_{j=p}^{q-1} |f(t_{j+1}) - f(t_j)|$$
$$\leq \bigvee_a^c (f) + \bigvee_c^b (f),$$
which yields
$$\bigvee_a^b (f) \leq \bigvee_a^c (f) + \bigvee_c^b (f). \tag{1.16}$$
Now, in view of (1.16) and (1.15), we get (1.14). This completes the proof. $\square$

Theorem 1.20. *A function f defined and finite on $[a, b]$ is a function of bounded variation on $[a, b]$ if and only if it is representable as the difference of two increasing functions.*

Proof. Suppose f is representable as the difference of two increasing functions f_1 and f_2 on $[a, b]$, i.e.,
$$f(t) = f_1(t) - f_2(t), \quad t \in [a, b].$$
Then, by Theorem 1.14, it follows that f_1 and f_2 are functions of bounded variation on $[a, b]$. Hence, applying Theorem 1.17, we conclude that f is a function of bounded variation on $[a, b]$. Conversely, suppose f is a function of bounded variation on $[a, b]$. Define function
$$g(t) = \begin{cases} \bigvee_a^t (f) & \text{for } t \in (a, b], \\ 0 & \text{for } t = a. \end{cases}$$
By Theorem 1.19, it follows that g is an increasing function on $[a, b]$. Consider the function
$$h(t) = g(t) - f(t), \quad t \in [a, b].$$

Now, assume that $a \leq t < s \leq b$. Let $t = a$. Then,
$$h(a) = g(a) - f(a)$$
$$= -f(a),$$
$$h(s) = g(s) - f(s)$$
$$= \bigvee_a^s(f) - f(s)$$
$$\geq f(s) - f(a) - f(s)$$
$$= -f(a)$$
$$= h(a).$$

Next, let $a < t$. Then,
$$h(s) - h(t) = g(s) - f(s) - g(t) + f(t)$$
$$= g(s) - g(t) - f(s) + f(t)$$
$$= \bigvee_a^s(f) - \bigvee_a^t(f) - f(s) + f(t)$$
$$= \bigvee_t^s(f) - f(s) + f(t)$$
$$\geq f(s) - f(t) - f(s) + f(t)$$
$$= 0.$$

Thus h is an increasing function on $[a, b]$ and
$$f(t) = g(t) - h(t), \quad t \in [a, b].$$
This completes the proof. □

Corollary 1.3. *If f is a function of bounded variation on $[a, b]$, then at almost every point of $[a, b]$, the derivative f' exists and is finite. Furthermore, f' is a summable function.*

Proof. Since f is of bounded finite variation on $[a, b]$, by Theorem 1.20, it follows that f can be represented in the form
$$f = f_1 - f_2,$$

where f_1 and f_2 are increasing functions on $[a,b]$. By Theorem 1.11, it follows that f_1 and f_2 have finite derivative at almost all points of $[a,b]$. By Theorem 1.12, we get that f_1' and f_2' are summable. This completes the proof. $\square$

Corollary 1.4. *The set of points of discontinuity of a function f of bounded variation on $[a,b]$ is at most denumerable. At every point t_0 of discontinuity, both $f(t_0+)$ and $f(t_0-)$ exist.*

Proof. Since f is of bounded variation on $[a,b]$, by Theorem 1.20, it follows that f can be represented in the form

$$f(t) = f_1(t) - f_2(t), \quad t \in [a,b],$$

where f_1 and f_2 are increasing functions on $[a,b]$. By Theorem 1.2, it follows that the sets of points of discontinuity of f_1 and f_2 are at most denumerable and $f_1(t_0+)$, $f_1(t_0-)$, $f_2(t_0+)$, and $f_2(t_0-)$ exist for any point t_0 of discontinuity of f_1 or f_2. This completes the proof. $\square$

Theorem 1.21. *A function of bounded variation can be written as the sum of its saltus function and a continuous function of bounded variation.*

Proof. Let f be a function of bounded variation on $[a,b]$ and the functions g and h be as in the proof of Theorem 1.20. Let the sequence $\{t_n\}_{n \in \mathbb{N}}$ consist of all points which are points of discontinuity of at least one of the functions g and h. Consider the saltus functions

$$S^g(t) = (g(a+) - g(a)) + \sum_{t_k < t}(g(t_k+) - g(t_k-)) + (g(t) - g(t-))$$

and

$$S^h(t) = (h(a+) - h(a)) + \sum_{t_k < t}(h(t_k+) - h(t_k-)) + (h(t) - h(t-)),$$

$a < t \leq b$. Let

$$S(t) = S^g(t) - S^h(t), \quad t \in [a,b].$$

Then,
$$S(t) = (g(a+) - g(a)) + \sum_{t_k < t}(g(t_k+) - g(t_k-)) + (g(t) - g(t-))$$
$$- (h(a+) - h(a)) + \sum_{t_k < t}(h(t_k+) - h(t_k-)) + (h(t) - h(t-))$$
$$= (f(a+) - f(a)) + \sum_{t_k < t}(f(t_k+) - f(t_k-)) + (f(t) - f(t-)),$$

$a < t \leq b$, $S(a) = 0$. Without any restriction, assume that all elements of $\{t_n\}_{n\in\mathbb{N}}$ are points of discontinuity of f. Then, by Theorem 1.3, it follows that $g - S^g$ and $h - S^h$ are increasing and continuous functions. Consequently, in view of Theorem 1.20, S is a continuous function of bounded variation on $[a,b]$. This completes the proof. $\square$

Example 1.4. Consider the Dirichlet function f defined in Example 1.3 on $[a,b]$, where $a,b \in \mathbb{Q}$. Take $n \in \mathbb{N}$ arbitrarily and
$$a = t_0 < t_1 < t_2 < \cdots < t_{n-1} < t_n = b.$$
such that $t_{2k-1} \in \mathbb{R} \setminus \mathbb{Q}$ and $t_{2k} \in \mathbb{Q}$, $k \in \mathbb{N}$. Then,
$$\bigvee_a^b(f) \geq |f(t_1) - f(t_0)| + |f(t_2) - f(xt_1)| + \cdots + |f(t_n) - f(t_{n-1})|$$
$$= |0 - 1| + |1 - 0| + \cdots + |1 - 0|$$
$$= 1 + \cdots + 1$$
$$= n.$$
Thus, f is not a function of bounded variation on $[a,b]$.

Example 1.5. Consider the function
$$f(t) = \begin{cases} \frac{1}{1-t} & \text{for } t \neq 1, \\ 0 & \text{for } t = 1 \end{cases}$$
on $[0,1]$. For $t \neq 1$, we have
$$f'(t) = \frac{1}{(1-t)^2}$$
$$> 0.$$

Thus, f is increasing on $(0,1)$ and is of bounded variation on any closed subinterval of $(0,1)$. Since f has a vertical asymptote at $t=1$, we can make the sum
$$\sum_{j=1}^{n} |f(t_j) - f(t_{j-1})|$$
as large as we like by choosing partition points close to 1. Thus, f is not a function of bounded variation on $[0,1]$.

Example 1.6. Consider the function
$$f(t) = \begin{cases} \sqrt[3]{t} \sin\left(\frac{\pi}{t}\right) & \text{for } t \neq 0, \\ 0 & \text{for } t = 0 \end{cases}$$
on $[0,1]$. Take $n \in \mathbb{N}$ even and
$$t_0 = 0, \quad t_n = 1, \quad t_{n-(2k+1)} = \frac{1}{k+3}, \quad t_{n-2k} = \frac{2}{2k+3}, \quad k \in \mathbb{N}.$$
Then,
$$\bigvee_0^1(f) \geq \sum_{j=1}^{n} |f(t_j) - f(t_{j-1})|$$
$$\geq \sum_{j=1}^{\frac{n}{2}} |f(t_{2j+1}) - f(t_{2j})|$$
$$= \sum_{k=1}^{n} |f(t_{n-(2k+1)}) - f(t_{n-2k})|$$
$$= \sum_{k=1}^{n} \left| f\left(\frac{1}{k+3}\right) - f\left(\frac{2}{2k+3}\right) \right|$$
$$= \sum_{k=1}^{n} \sqrt[3]{\frac{2}{2k+3}}.$$
Since the series
$$\sum_{k=1}^{\infty} \sqrt[3]{\frac{2}{2k+3}}$$
is divergent, we conclude that f is not of bounded variation on $[0,1]$.

Exercise 1.3. Prove that the function

$$f(t) = \begin{cases} t \cos\left(\frac{\pi}{2t}\right) & \text{for } t \in (0, 1], \\ 0 & \text{for } t = 0 \end{cases}$$

is not a function of bounded variation on $[0, 1]$.

1.4 The Helly Principle of Choice

Theorem 1.22 (The Helley Selection Principle). *Let $H = \{f\}$ be an infinite family of functions defined on $[a, b]$ such that*

$$|f(t)| \leq K, \quad t \in [a, b],$$

for any $f \in H$ and for some $K > 0$. Then, for any denumerable set $E \subset [a, b]$, there exists a sequence $\{f_n\}_{n \in \mathbb{N}} \subset H$ that converges at every point of the set E.

Proof. Let $\{t_k\}_{k \in \mathbb{N}} \subset [a, b]$. Consider the set

$$\{f(t_1) \colon f \in H\}.$$

This set is bounded. Then, applying the Bolzano–Weierstrass theorem, it follows that there exists a sequence $\{f_n^{(1)}(t_1)\}_{n \in \mathbb{N}} \subset H$ such that

$$\lim_{n \to \infty} f_n^{(1)}(t_1) = A_1.$$

Next, consider the sequence $\{f_n^{(2)}(t_2)\}_{n \in \mathbb{N}}$, which is also bounded. Applying the Bolzano–Weierstrass theorem, we get a sequence $\{f_n^{(2)}(t_2)\}_{n \in \mathbb{N}} \subset \{f_n^{(1)}(t_2)\}_{n \in \mathbb{N}}$ so that

$$\lim_{n \to \infty} f_n^{(2)}(t_2) = A_2.$$

Continuing this process, we construct a denumerable set of convergent sequences

$$\left\{\{f_n^{(1)}(t_1)\}_{n \in \mathbb{N}}, \{f_n^{(2)}(t_1)\}_{n \in \mathbb{N}}, \ldots, \{f_n^{(k)}(t_1)\}_{n \in \mathbb{N}}\right\}$$

such that

$$\lim_{n\to\infty} f_n^{(1)}(t_1) = A_1,$$

$$\lim_{n\to\infty} f_n^{(2)}(t_2) = A_2,$$

$$\vdots$$

$$\lim_{n\to\infty} f_n^{(k)}(t_k) = A_k,$$

and so on. Now, we form the sequence $\{f_n^{(n)}\}_{n\in\mathbb{N}}$. For any fixed $k \in \mathbb{N}$, we have

$$\{f_n^{(n)}(t_n)\}_{n\in\mathbb{N}} \subset \{f_n^{(k)}(t_k)\}.$$

Then the formed sequence $\{f_n^{(n)}\}_{n\in\mathbb{N}}$ is convergent at each point of E. This completes the proof. □

Theorem 1.23. *Let $F = \{f\}$ be an infinite family of increasing functions defined on $[a,b]$ such that*

$$|f(t)| \leq K, \quad t \in [a,b],$$

for any $f \in F$ and for some $K > 0$. Then there is a sequence $\{f_n\}_{n\in\mathbb{N}} \subset F$ of increasing functions that converges, at every point of the interval $[a,b]$, to an increasing function ϕ on $[a,b]$.

Proof. Let E be the set of all rational numbers in the interval $[a,b]$ including a provided it is rational. We apply Theorem 1.22 and we find a sequence $F_0 = \{f^{(n)}\}_{n\in\mathbb{N}}$ such that $\lim_{n\to\infty} f^{(n)}(t_k)$ exists for any $t_k \in E$. Define a function ψ as

$$\psi(t_k) = \lim_{n\to\infty} f^{(n)}(t_k), \quad t_k \in E.$$

Note that ψ is an increasing function on E. For $t \in [a,b] \setminus E$, define

$$\psi(t) = \sup_{t_k < t} \{\psi(t_k)\}.$$

Then, ψ is an increasing function on $[a,b]$ and the set of points of discontinuity of ψ is at most denumerable. Suppose ψ is continuous

at point $t_0 \in [a,b]$. Take $\varepsilon > 0$ arbitrarily. Let $t_k, t_l \in E$ be such that
$$t_k < t_0 < t_l \quad \text{and} \quad \psi(t_l) - \psi(t_k) < \frac{\varepsilon}{2}.$$
Now, we take $N \in \mathbb{N}$ so that
$$|f^{(n)}(t_k) - \psi(t_k)| < \frac{\varepsilon}{2},$$
$$|f^{(n)}(t_l) - \psi(t_l)| < \frac{\varepsilon}{2}, \quad n > N.$$
Then
$$f^{(n)}(t_k) \leq f^{(n)}(t_0) \leq f^{(n)}(t_l), \quad n > N,$$
and
$$\psi(t_l) - \frac{\varepsilon}{2} < f^{(n)}(t_l) < \psi(t_l) + \frac{\varepsilon}{2},$$
$$\psi(t_k) - \frac{\varepsilon}{2} < f^{(n)}(t_k) < \psi(t_k) + \frac{\varepsilon}{2}, \quad n > N,$$
and
$$\psi(t_0) - \psi(t_k) < \frac{\varepsilon}{2},$$
$$\psi(t_l) - \psi(t_0) < \frac{\varepsilon}{2}.$$
Hence
$$\psi(t_k) > \psi(t_0) - \frac{\varepsilon}{2},$$
$$\psi(t_l) < \psi(t_0) + \frac{\varepsilon}{2},$$
and
$$\psi(t_0) - \varepsilon = \psi(t_0) - \frac{\varepsilon}{2} - \frac{\varepsilon}{2}$$
$$< \psi(t_k) - \frac{\varepsilon}{2}$$
$$< f^{(n)}(t_k)$$
$$< f^{(n)}(t_0)$$

$$< f^{(n)}(t_l)$$
$$< \psi(t_l) + \frac{\varepsilon}{2}$$
$$< \psi(t_0) + \frac{\varepsilon}{2} + \frac{\varepsilon}{2}$$
$$= \psi(t_0) + \varepsilon,$$

whereupon we get the chain

$$\psi(t_0) - \varepsilon < f^{(n)}(t_k) < f^{(n)}(t_0) < f^{(n)}(t_l) < \psi(t_0) + \varepsilon, \quad n > N,$$

and

$$\lim_{n \to \infty} f^{(n)}(t_0) = \psi(t_0).$$

Since t_0 was arbitrarily chosen continuity point of ψ, we conclude that

$$\lim_{n \to \infty} f^{(n)}(t) = \psi(t) \qquad (1.17)$$

holds at every point $t \in [a, b]$, at which ψ is continuous. Let Q denote the set of all points of discontinuity of ψ lying in $[a, b]$. Note that Q is at most denumerable. Now, we apply Theorem 1.22 for F_0 and Q, and we find a sequence $\{f_n\}_{n \in \mathbb{N}} \subset F$ which converges at any point of Q. Since $\{f^{(n)}\}_{n \in \mathbb{N}}$ converges at any point of $[a, b] \setminus Q$, we conclude that $\{f_n\}_{n \in \mathbb{N}}$ converges at any point of $[a, b]$. We set

$$\phi(t) = \lim_{n \to \infty} f_n(t) \quad \text{for } t \in [a, b],$$

and this completes the proof. □

Theorem 1.24 (The Helley first theorem). *Let $F = \{f\}$ be an infinite family of functions defined on $[a, b]$ such that*

$$|f(t)| \leq K, \quad \bigvee_a^b (f) \leq K, \quad t \in [a, b],$$

for any $f \in F$ and for some $K > 0$. Then, there exists a sequence $\{f_n\}_{n \in \mathbb{N}} \subset F$ that converges at every point of $[a, b]$ to a function ϕ of bounded variation on $[a, b]$.

Proof. For $f \in F$, set

$$\pi_f(t) = \bigvee_a^t (f)$$

and

$$\nu_f(t) = \pi_f(t) - f(t), \quad t \in [a, b].$$

Note that π_f and ν_f are increasing functions on $[a, b]$ for any $f \in F$. Further,

$$|\pi_f(t)| \leq K$$

and

$$|\nu_f(t)| \leq 2K, \quad t \in [a, b],$$

for any $f \in F$. Now, we apply Theorem 1.23 to the family

$$\{\pi_f \colon f \in H\},$$

and we find a sequence $\{\pi_{f_n}\}_{n \in \mathbb{N}}$ that converges to a function α at any point of $[a, b]$. Thus, we get the sequence $\{\nu_{f_n}\}_{n \in \mathbb{N}}$ extending it to the sequence $\{f_n\}_{n \in \mathbb{N}} \subset F$. We apply Theorem 1.23 and find a sequence $\{\pi_{f_{n_m}}\}_{m \in \mathbb{N}}$ so that

$$\lim_{m \to \infty} \pi_{f_{n_m}}(t) = \beta(t), \quad t \in [a, b].$$

The sequence

$$\{f_{n_m} = \pi_{n_m} - \nu_{n_m}\}_{m \in \mathbb{N}} \subset F$$

converges to $\alpha - \beta$ at any point of $[a, b]$. This completes the proof. $\square$

1.5 Continuous Functions of Bounded Variation

Theorem 1.25. *Let f be a function of bounded variation defined on $[a, b]$. If f is continuous at $t_0 \in [a, b]$, then the function*

$$g(t) = \bigvee_a^t (f), \quad t \in [a, b],$$

is continuous at t_0.

Proof. Suppose that $t_0 < b$. Take $\varepsilon > 0$ arbitrarily and the points
$$t_0 < t_1 < \cdots < t_n < b,$$
such that
$$\sum_{j=1}^{n-1} |f(t_{j+1}) - f(t_j)| > \bigvee_{t_0}^{b}(f) - \varepsilon$$
and
$$|f(t_1) - f(t_0)| < \varepsilon.$$
Then,
$$\bigvee_{t_0}^{b}(f) < \varepsilon + \sum_{j=0}^{n-1} |f(t_{j+1}) - f(t_j)|$$
$$= \varepsilon + |f(t_1) - f(t_0)| + \sum_{j=1}^{n-1} |f(t_{j+1}) - f(t_j)|$$
$$< 2\varepsilon + \sum_{j=1}^{n-1} |f(t_{j+1}) - f(t_j)|$$
$$\leq 2\varepsilon + \bigvee_{t_1}^{b}(f),$$
whereupon
$$\bigvee_{t_0}^{t_1}(f) < 2\varepsilon$$
and
$$|g(t_1) - g(t_0)| < 2\varepsilon.$$
This implies that
$$|g(t_0+) - g(t_0)| < \varepsilon.$$
Since $\varepsilon > 0$ was arbitrarily chosen, we conclude that $g(t_0+) = g(t_0)$. As above, it can be proved that $g(t_0-) = g(t_0)$ if $t_0 > a$. This completes the proof. □

The Stieltjes Integral

Corollary 1.5. *A continuous function f of bounded variation on $[a, b]$ can be written as the difference of two continuous increasing functions.*

Proof. By Theorem 1.20 and its proof, it follows that f can be written in the form
$$f(t) = g(t) - h(t), \quad t \in [a, b],$$
where g and h are increasing functions such that
$$g(t) = \bigvee_a^t (f), \quad t \in [a, b].$$
By Theorem 1.25, we have that g is a continuous function on $[a, b]$. Then, h is a continuous function on $[a, b]$. This completes the proof. $\square$

For any partition P, given by (1.13), denote
$$\mu(P) = \max_{k \in \{0,\ldots,n-1\}} (t_{k+1} - t_k).$$

Denote $\mathscr{C}([a,b])$, the set of continuous functions on $[a,b]$.

Let $f \in \mathscr{C}([a,b])$ and $\omega_k(f, P)$, $k \in \{0, \ldots, n-1\}$, denote the oscillation of f on $[t_k, t_{k+1}]$, $k \in \{0, \ldots, n-1\}$, i.e., $\omega_k(f, P) = \sup\{|f(u) - f(v)| : u, v \in [t_k, t_{k+1}]\}$, and
$$\Omega(f, P) = \sum_{k=0}^{n-1} \omega_k(f, P).$$

Theorem 1.26. *Let $f \in \mathscr{C}([a,b])$. Then,*
$$\lim_{\mu(P) \to 0} V(f, P) = \bigvee_a^b (f) \quad \text{and} \quad \lim_{\mu(P) \to 0} \Omega(f, P) = \bigvee_a^b (f).$$

Proof. Consider the partition P given by (1.13). Let
$$Q: a = t_0 < t_1 < \cdots < t_l < s < t_{l+1} < \cdots < t_n = b$$

be another partition of $[a,b]$. Then,

$$V(f,Q) = \sum_{j=0}^{l-1} |f(t_{j+1}) - f(t_j)| + |f(s) - f(t_l)|$$

$$+ |f(t_{l+1}) - f(s)| + \sum_{j=l+1}^{n-1} |f(t_{j+1}) - f(t_j)|$$

$$\geq \sum_{j=0}^{l-1} |f(t_{j+1}) - f(t_j)| + |f(t_{l+1}) - f(t_l)|$$

$$+ \sum_{j=l+1}^{n-1} |f(t_{j+1}) - f(t_j)|$$

$$= V(f,P).$$

From here, we conclude that $V(f,\cdot)$ does not decrease when new points of subdivision are added. Note that

$$\omega_l(f,P) = \sup_{t \in [t_l, t_{l+1}]} f(t) - \inf_{t \in [t_l, t_{l+1}]} f(t)$$

$$\geq |f(s) - f(t_l)|$$

and

$$\omega_l(f,P) \geq |f(t_{l+1}) - f(s)|.$$

Thus, if we add a new point y in P that falls in $[t_l, t_{l+1}]$, then the increase in the sum $V(f,Q)$ due to this point is not greater than twice the oscillation ω_l of f on $[t_l, t_{l+1}]$. Let

$$A < \bigvee_a^b (f)$$

and find a partition $R: a = t_0^* < t_1^* < \cdots < t_{m-1}^* < t_m^* = b$ of $[a,b]$ so that

$$V(f,R) > A. \tag{1.18}$$

Since $f \in \mathscr{C}([a,b])$, there is a $\delta > 0$ small enough so that

$$|f(u) - f(v)| < \frac{V(f,R) - A}{4m},$$

provided that
$$|u - v| < \delta.$$
Let the partition P, given by (1.13), be such that $\mu(P) < \delta$. Also, let $S = P \cup R$. Then,
$$V(f, S) \geq V(f, R).$$
Note that S is obtained from R by the repetition of the process of adding a single point. Each addition of a point of subdivision increases the sum $V(f, R)$ by a number that is less than
$$\frac{V(f, R) - A}{2m}.$$
Then,
$$V(f, S) - V(f, P) < \frac{V(f, R) - A}{2}.$$
Hence,
$$V(f, P) \geq V(f, S) - \frac{V(f, R) - A}{2}$$
$$\geq V(f, R) - \frac{V(f, R) - A}{2}$$
$$= \frac{A + V(f, R)}{2}.$$
Therefore, (1.18) holds for any $\lambda(P) < \delta$. Since
$$V(f, P) \leq \bigvee_a^b (f)$$
by all $V(f, P)$, it follows that
$$\lim_{\lambda(P) \to 0} V(f, P) = \bigvee_a^b (f).$$
Now, we observe that
$$\Omega(f, P) \geq V(f, P). \tag{1.19}$$

Note that if we find a sum $\Omega(f,\cdot)$ corresponding to any method of subdivision and then add new points at which f takes its maximum and minimum on $[x_k, x_{k+1}]$, $k \in \{0,\ldots,n-1\}$, then the sum $V(f,\cdot)$ corresponding to the method of subdivision will be not less that $\Omega(f,\cdot)$. Therefore,

$$\Omega(f,\cdot) \leq \bigvee_a^b (f).$$

From here, applying (1.19), we get

$$\lim_{\mu(P) \to 0} \Omega(f,P) = \bigvee_a^b (f).$$

This completes the proof. $\square$

Definition 1.7. Let $f \in \mathscr{C}([a,b])$. Then there exist

$$m = \min_{t \in [a,b]} f(t) \quad \text{and} \quad M = \max_{t \in [a,b]} f(t).$$

For $s \in [m, M]$, the Banach indicatrix of s is defined to be the number that is equal to the number of roots of the equation $f(t) = s$, $t \in [a,b]$. It is denoted by $N(s)$. Note that the Banach indicatrix is an integer-valued function.

Theorem 1.27. *The Banach indicatrix is measurable and*

$$\bigvee_m^M (N(s)) = \bigvee_a^b (f).$$

Proof. Let $n \in \mathbb{N}$ and

$$d_1 = \left[a, a + \frac{b-a}{2^n}\right],$$

$$d_k = \left(a + (k-1)\frac{b-a}{2^n}, a + k\frac{b-a}{2^n}\right], \quad k \in \{2, 3, \ldots, 2^n\}.$$

Let $s \in [m, M]$ and consider the equation

$$f(t) = s \quad \text{for } t \in [a,b]. \tag{1.20}$$

For $k \in \{1, \ldots, 2^n\}$, define

$$L_k(s) = \begin{cases} 0 & \text{if } f(t) = s \text{ has at least one root in the interval } d_k, \\ 1 & \text{if } f(t) = s \text{ has no root in the interval } d_k. \end{cases}$$

Also, let

$$m_k = \inf_{t \in d_k} f(t) \quad \text{and} \quad M_k = \sup_{t \in d_k} f(t) \quad \text{for } k \in \{1, \ldots, 2^n\}.$$

Then

$$L_k(s) = \begin{cases} 1 & \text{for } s \in (m_k, M_k), \\ 0 & \text{for } s \in [a,b] \setminus (m_k, M_k), \end{cases} \quad k \in \{1, \ldots, 2^n\},$$

and

$$\int_m^M L_k(s) ds = \int_{m_k}^{M_k} L_k(s) ds$$
$$= \int_{m_k}^{M_k} ds$$
$$= M_k - m_k$$
$$= \omega_k, \quad k \in \{1, \ldots, 2^n\}.$$

Define

$$N_n(s) = L_1(s) + \cdots + L_{2^n}(s), \quad s \in [m, M].$$

Observe that $N_n(s)$, $s \in [m, M]$, equals the number of those intervals d_k, $k \in \{1, \ldots, 2^n\}$, which contain at least one root of (1.20). Note that N_n is a measurable function on $[m, M]$ and

$$\int_m^M N_n(s) ds = \sum_{k=1}^{2^n} \omega_k.$$

Now, applying Theorem 1.26, we get

$$\lim_{n \to \infty} \int_m^M N_n(s) ds = \bigvee_a^b (f).$$

Note that
$$N_1(s) \leq N_2(s) \leq \ldots.$$
Hence, finite or infinite, there exists
$$N^*(s) = \lim_{n\to\infty} N_n(s), \quad s \in [m, M],$$
as a measurable function. Applying the Levi theorem, we find
$$\int_m^M N^*(s)ds = \lim_{n\to\infty} \int_m^M N_n(s)ds$$
$$= \bigvee_a^b (f).$$
Note that
$$N_n(s) \leq N(s), \quad n \in \mathbb{N}, \ s \in [m, M],$$
and then
$$N^*(s) \leq N(s), \quad s \in [m, M]. \tag{1.21}$$
Let $q \in \mathbb{N}$ be such that $q \leq N(s)$. Then, we can find q distinct roots
$$t_1 < t_2 < \cdots < t_q$$
of (1.20). Take $n \in \mathbb{N}$ large enough so that
$$\frac{b-a}{2^n} < \min_{k\in\{1,\ldots,q-1\}} (t_{k+1} - t_k).$$
Then all q roots t_k, $k \in \{1, \ldots, q\}$, will fall into distinct intervals d_k, $k \in \{1, \ldots, q\}$, so that
$$N_n(s) \geq q, \quad s \in [m, M],$$
and therefore,
$$N^*(s) \geq q, \quad s \in [m, M].$$
If $N(s) = \infty$, $s \in [m, M]$, then we can take q arbitrary large so that $N^*(s) = \infty$, $s \in [m, M]$. If $N(s) < \infty$, $y \in [m, M]$, then we take $q = N(s)$, $s \in [m, M]$, and
$$N^*(s) \geq N(s), \quad s \in [m, M].$$

Hence, in view of (1.21), we obtain

$$N^*(s) = N(s), \quad s \in [m, M].$$

This completes the proof. □

Corollary 1.6. *A continuous function f is of bounded variation if and only if its Banach indicatrix N is summable.*

1.6 The Stieltjes Integral

Let the functions f and g be finite and defined on the closed interval $[a, b]$. Consider the partition P given by (1.13), and choose tag points $\xi_k \in [x_k, x_{k+1}]$, $k \in \{0, \ldots, n-1\}$. Form the sum

$$\sigma(P, f, g) = \sum_{k=0}^{n-1} f(\xi_k)(g(t_{k+1}) - g(t_k)).$$

Definition 1.8. If there exists a finite limit $\mathscr{S}$ such that

$$\mathscr{S} = \lim_{\mu(P) \to 0} \sigma(P, f, g),$$

then this finite limit is said to be the Stieltjes integral of f with respect to g and it is denoted by

$$\int_a^b f(t) d_g t.$$

The exact meaning of this definition is as follows. The number $\mathscr{S}$ is said to be the Stieltjes integral of f with respect to g provided for any $\varepsilon > 0$, there exists a $\delta = \delta(\varepsilon) > 0$ such that for any partition P for which $\mu(P) < \delta$ the inequality

$$|\sigma(P, f, g) - \mathscr{S}| < \varepsilon$$

holds for any choice of the tag points ξ_k, $k \in \{0, \ldots, n-1\}$.

Remark 1.3. Note that the Riemann integral is a special case of the Stieltjes integral obtained by setting $g(t) = t$, $t \in [a, b]$.

Example 1.7. Consider the functions
$$f(t) = \begin{cases} 0 & \text{for } t \in [-1,0], \\ 1 & \text{for } t \in (0,1], \end{cases}$$
and
$$g(t) = \begin{cases} 0 & \text{for } t \in [-1,0), \\ 1 & \text{for } t \in [0,1]. \end{cases}$$
We will compute
$$\mathscr{S}_1 = \int_{-1}^{0} f(t)\,d_g t,$$
$$\mathscr{S}_2 = \int_{0}^{1} f(t)\,d_g t,$$
and
$$\mathscr{S}_3 = \int_{-1}^{1} 2f(t)\,d_g t.$$
Consider the partitions P_1 of $[-1,0]$ and P_2 of $[0,1]$ given by
$$P_1: -1 = t_0 < t_1 < \cdots < t_m = 0 \quad \text{and}$$
$$P_2: 0 = s_0 < s_1 < \cdots < s_n = 1, \quad \text{respectively.}$$
Take $\xi_k \in [t_k, t_{k+1}]$, $k \in \{0, \ldots, m-1\}$ and $\eta_k \in [s_k, s_{k+1}]$, $k \in \{0, \ldots, n-1\}$. Then,
$$\sigma_1(P_1, f, g) = \sum_{k=0}^{m-1} f(\xi_k)(g(t_{k+1}) - g(t_k))$$
$$= 0$$
and
$$\sigma_2(P_2, f, g) = \sum_{k=0}^{n-1} f(\eta_k)(g(s_{k+1}) - g(s_k))$$
$$= 0,$$
whereupon
$$\mathscr{S}_1 = 0 \quad \text{and}$$
$$\mathscr{S}_2 = 0.$$

Now, let
$$P_3\colon\ -1 = u_0 < u_1 < \cdots < u_i < 0 < u_{i+1} < \cdots < u_p = 1$$
be a partition of $[-1, 1]$ and $\zeta_k \in [u_k, u_{k+1}]$, $k \in \{0, \ldots, p-1\}$. Then,
$$\sigma(P_3, f, g) = \sum_{k=0}^{p-1} f(\zeta_k)(g(u_{k+1}) - g(u_k))$$
$$= f(\zeta_i)(g(u_{i+1}) - g(u_i))$$
$$= f(\zeta_i)$$
$$= \begin{cases} 0 & \text{if } \zeta_i \in [-1, 0], \\ 1 & \text{if } \zeta_i \in (0, 1]. \end{cases}$$

Thus, $\mathscr{S}_3$ does not exist.

Exercise 1.4. Let
$$f(t) = \sin t$$
and
$$g(t) = t^3, \quad t \in [0, \pi].$$
Compute
$$\int_0^1 f(t) d_g t.$$

Answer 1.1. 2π.

We list the following obvious properties of the Stieltjes integral, which we leave to the reader as an exercise.

Theorem 1.28. *Let the functions f, f_1, f_2, g, g_1, g_2 be finite and defined on $[a, b]$.*

(1) *Suppose that the integrals*
$$\int_a^b f_1(t) d_g t \quad \text{and} \quad \int_a^b f_2(t) d_g t$$
exist. Then,
$$\int_a^b (f_1(t) + f_2(x)) d_g t = \int_a^b f_1(t) d_g t + \int_a^b f_2(t) d_g t.$$

(2) *Suppose that the integrals*

$$\int_a^b f(t)dg_1 t \quad \text{and} \quad \int_a^b f(t)dg_2 t$$

exist. Then,

$$\int_a^b f(t)dg_{1+g_2}t = \int_a^b f(t)dg_1 t + \int_a^b f(t)dg_2 t.$$

(3) *Suppose that the integral*

$$\int_a^b f(t)dg t$$

exist. Then,

$$\int_a^b kf(t)d_l g t = k\int_a^b f(t)dg t$$

for any $k, l \in \mathbb{R}$.

(4) *Suppose that $a < c < b$ and the integrals*

$$\int_a^c f(t)dg t, \quad \int_c^b f(t)dg t, \quad \text{and} \quad \int_a^b f(t)dg t$$

exist. Then,

$$\int_a^b f(t)dg t = \int_a^c f(t)dg t + \int_c^b f(t)dg t. \qquad (1.22)$$

Remark 1.4. Let f and g be as in Example 1.7. Also, let $a = -1$, $c = 0$, $b = 1$. By Example 1.7, we have that $\mathscr{S}_1$ and $\mathscr{S}_2$ exist, but $\mathscr{S}_3$ does not exist. Therefore, (1.22) does not hold.

In the following, we establish the formula for integration by parts for the Stieltjes integral.

Theorem 1.29. *Let f and g be finite and defined on $[a, b]$. Then, the existence of one of the integrals*

$$\int_a^b f(t)dg t \quad \text{and} \quad \int_a^b g(t)df t$$

implies the existence of the other. In this case, the equality

$$\int_a^b f(t)dg t + \int_a^b g(t)df t = f(b)g(b) - f(a)g(a) \qquad (1.23)$$

holds.

Proof. Suppose that the integral

$$\int_a^b f(t)d_g t$$

exists. Consider the partition P of $[a,b]$ given by (1.13) and let $\xi_k \in [x_k, x_{k+1}]$, $k \in \{0, \ldots, n-1\}$. Then,

$$\sigma(P, f, g) = \sum_{k=0}^{n-1} f(\xi_k)(g(t_{k+1}) - g(t_k))$$

$$= \sum_{k=0}^{n-1} f(\xi_k)g(t_{k+1}) - \sum_{k=0}^{n-1} f(\xi_k)g(t_k)$$

$$= \sum_{k=1}^{n} f(\xi_{k-1})g(t_k) - \sum_{k=0}^{n-1} f(\xi_k)g(t_k)$$

$$= \sum_{k=1}^{n-1} f(\xi_{k-1})g(t_k) + f(\xi_{n-1})g(t_n)$$

$$- \sum_{k=1}^{n-1} f(\xi_k)g(t_k) - f(\xi_0)g(t_0)$$

$$= -\sum_{k=1}^{n-1} (f(\xi_k) - f(\xi_{k-1}))g(t_k) + f(\xi_{n-1})g(t_n) - f(\xi_0)g(t_0)$$

$$= f(b)g(b) - f(a)g(a) - f(b)g(b) + f(a)g(a)$$

$$- \sum_{k=1}^{n-1} (f(\xi_k) - f(\xi_{k-1}))g(t_k) + f(\xi_{n-1})g(t_n) - f(\xi_0)g(t_0)$$

$$= f(b)g(b) - f(a)g(a) - (f(\xi_0) - f(a))g(a)$$

$$- \sum_{k=1}^{n-1} (f(\xi_k) - f(\xi_{k-1}))g(t_k) - (f(b) - f(\xi_{n-1}))g(b)$$

$$= f(b)g(b) - f(a)g(a) - \sigma(P_1, g, f),$$

where

$$P_1 \colon a \leq \xi_0 \leq \xi_1 \leq \cdots \leq \xi_{n-1} \leq b,$$

i.e.,
$$\sigma(P, f, g) = f(b)g(b) - f(a)g(a) - \sigma(P_1, g, f). \tag{1.24}$$

Note that $t_0 \in [a, \xi_0]$, $t_1 \in [\xi_0, \xi_1], \ldots, t_{n-1} \in [\xi_{n-2}, \xi_{n-1}]$, $t_n \in [\xi_{n-1}, b]$. Moreover, $\mu(P) \to 0$ if and only if $\mu(P_1) \to 0$. Hence, applying (1.24), we find

$$\int_a^b f(t)d_g t = f(b)g(b) - f(a)g(a) - \int_a^b g(t)d_f t.$$

This completes the proof. □

Theorem 1.30. *Let f and g be finite and defined on $[a, b]$. Then the integral*

$$\int_a^b f(t)d_g t$$

exists provided $f \in \mathscr{C}([a, b])$ and g is of bounded variation on $[a, b]$.

Proof. Without any restriction, assume that g is increasing, because by Theorem 1.20, it follows that it is the difference of two increasing functions. Consider the partition P given by (1.13). Let

$$m_k = \min_{t \in [t_k, t_{k+1}]} f(t) \quad \text{and} \quad M_k = \max_{t \in [t_k, t_{k+1}]} f(t), \quad k \in \{1, \ldots, n-1\}.$$

Consider

$$s(P, f, g) = \sum_{k=0}^{n-1} m_k(g(t_{k+1}) - g(t_k))$$

and

$$S(P, f, g) = \sum_{k=0}^{n-1} M_k(g(t_{k+1}) - g(t_k)).$$

Note that

$$s(P, f, g) \leq \sigma(P, f, g) \leq S(P, f, g)$$

for any choices of tag points $\xi_k \in [t_k, t_{k+1}]$, $k \in \{0, \ldots, n-1\}$. Moreover, $s(\cdot, f, g)$ does not decrease and $S(\cdot, f, g)$ does not increase when

new points of subdivision are added. Let
$$\mathscr{S} = \sup\{s(\cdot, f, g)\}.$$
We have
$$s(\cdot, f, g) \leq \mathscr{S} \leq S(\cdot, f, g).$$
Then,
$$|\sigma(\cdot, f, g) - I| < S(\cdot, f, g) - s(\cdot, f, g).$$
Take $\varepsilon > 0$ arbitrarily. Then, there is $\delta = \delta(\varepsilon) > 0$ such that
$$|f(t) - f(s)| < \varepsilon$$
whenever $|t - s| < \delta$. Then if $\mu(P) < \delta$, we have
$$M_k - m_k < \varepsilon, \quad k \in \{0, \ldots, n-1\},$$
and
$$S(P, f, g) - s(P, f, g) = \sum_{k=0}^{n-1} (M_k - m_k)(g(t_{k+1}) - g(t_k))$$
$$< \varepsilon \sum_{k=0}^{n-1} (g(t_{k+1}) - g(t_k))$$
$$= \varepsilon(g(t_1) - g(a) + g(t_2) - g(t_1) + \cdots$$
$$+ g(b) - g(t_{n-1}))$$
$$= \varepsilon(g(b) - g(a)).$$
Therefore,
$$|\sigma(\cdot, f, g) - \mathscr{S}| < \varepsilon(g(b) - g(a)) \quad \text{provided} \quad \mu(\cdot) < \delta.$$
Consequently,
$$\lim_{\mu(P) \to 0} \sigma(P, f, g) = \mathscr{S},$$
i.e.,
$$\mathscr{S} = \int_a^b f(t) d_g t.$$
This completes the proof. □

Theorem 1.31. Let $f \in \mathscr{C}([a,b])$ and g be Riemann integrable over $[a,b]$ such that g' exists at every point of $[a,b]$. Then,

$$\int_a^b f(t)d_g t = \int_a^b f(t)g'(t)dt.$$

Proof. Since g' exists at every point of $[a,b]$, g satisfies the Lipschitz condition on $[a,b]$. By Theorem 1.15, it follows that g is of bounded variation on $[a,b]$. Now, applying Theorem 1.30, we conclude that the Stieltjes integral

$$\int_a^b f(t)d_g t$$

exists. Moreover, the product fg' is continuous almost everywhere on $[a,b]$, and thus, the Riemann integral

$$\int_a^b f(t)g'(t)dt$$

exists. Let us consider the partition P of $[a,b]$ given by (1.13). By the mean value theorem, it follows that there exists $\xi_k \in [t_k, t_{k+1}]$, $k \in \{0, \ldots, n-1\}$ such that

$$g(t_{k+1}) - g(t_k) = g'(\xi_k)(t_{k+1} - t_k), \quad k \in \{0, \ldots, n-1\}.$$

Then,

$$\sigma(P, f, g) = \sum_{k=0}^{n-1} f(\xi_k)(g(t_{k+1}) - g(t_k))$$

$$= \sum_{k=0}^{n-1} f(\xi_k)g'(\xi_k)(t_{k+1} - t_k),$$

whereupon

$$\int_a^b f(t)d_g t = \lim_{\lambda(P) \to 0} \sigma(P, f, g)$$

$$= \lim_{\lambda(P) \to 0} \sum_{k=0}^{n-1} f(\xi_k)g'(\xi_k)(t_{k+1} - t_k)$$

$$= \int_a^b f(t)g'(t)dt.$$

This completes the proof. $\square$

Example 1.8. Let
$$f(t) = e^t$$
and
$$g(t) = t^3, \quad t \in [0, 1].$$
Note that $g \in \mathscr{C}^1([0,1])$. Then, applying Theorem 1.31, we find
$$\int_0^1 f(t)d_g t = \int_0^1 f(t)g'(t)dt$$
$$= 3\int_0^1 t^2 e^t dt$$
$$= 3\left(t^2 e^t \Big|_{t=0}^{t=1} - 2\int_0^1 te^t dt\right)$$
$$= 3\left(e - 2te^t \Big|_{t=0}^{t=1} + 2\int_0^1 e^t dt\right)$$
$$= 3\left(e - 2e + 2e^t \Big|_{t=0}^{t=1}\right)$$
$$= 3(e - 2e + 2e - 2)$$
$$= 3(e - 2).$$

Exercise 1.5. Let
$$f(t) = t^4$$
and
$$g(t) = t^2, \quad t \in [0, 1].$$
Using Theorem 1.31, compute the integral
$$\int_0^1 f(t)d_g t.$$

Answer 1.2. $\frac{1}{3}$.

Theorem 1.32. Let $f \in \mathscr{C}([a,b])$ and g be a constant function on each of the intervals $(a, c_1), (c_1, c_2), \ldots, (c_m, b)$, where

$$a < c_1 < \cdots < c_m < b.$$

Then,

$$\int_a^b f(t) d_g t = f(a)(g(a+) - g(a)) + \sum_{k=1}^m f(c_k)(g(c_k+) - g(c_k-))$$
$$+ f(b)(g(b) - g(b-)).$$
(1.25)

Proof. Note that

$$\bigvee_a^b (g) = (g(a+0) - g(a)) + \sum_{k=1}^m (g(c_k+) - g(c_k-)) + (g(b) - g(b-))$$

$$< \infty.$$

Thus, g is of bounded variation on $[a, b]$. By Theorem 1.30, it follows that there exists the Stieltjes integral

$$\int_a^b f(t) d_g t$$

and

$$\int_a^b f(t) d_g t = \sum_{k=0}^m \int_{c_k}^{c_{k+1}} f(t) d_g t,$$

where $c_0 = a$ and $c_m = b$. Let

$$P_1 : c_k = s_0 < s_1 < \cdots < s_{p-1} < s_k = c_{k+1}, \quad k \in \{0, \ldots, m\}$$

be a partition of $[c_k, c_{k+1}]$ and $\xi_l \in [s_l, s_{l+1}]$, $l \in \{0, \ldots, p-1\}$. Then,

$$\sigma(P_1, f, g) = f(\xi_0)(g(c_k+) - g(c_k)) + f(\xi_{p-1})(g(c_{k+1}) - g(c_{k+1}-)),$$

$k \in \{0, \ldots, m\}$, and

$$\int_{c_k}^{c_{k+1}} f(t) d_g t = f(c_k)(g(c_k+) - g(c_k)) + f(c_{k+1})(g(c_{k+1}) - g(c_{k+1}-)),$$

$k \in \{0, \ldots, m\}$. Hence, we get (1.25). This completes the proof. □

Example 1.9. Let $f(t) = t^2 - t$, $t \in [0, 3]$ and

$$g(t) = \begin{cases} -1 & \text{for } t = 0, \\ 1 & \text{for } t \in (0, 1), \\ -2 & \text{for } t \in (1, 2), \\ 0 & \text{for } t \in (2, 3), \\ 1 & \text{for } t = 3. \end{cases}$$

We will find

$$\int_0^3 f(t) d_g t.$$

Here,

$$a = 0,$$
$$b = 3,$$
$$c_1 = 1,$$
$$c_2 = 2,$$
$$f(0) = 0,$$
$$f(1) = 1^2 - 1$$
$$= 0,$$
$$f(2) = 2^2 - 2$$
$$= 2,$$
$$f(3) = 3^2 - 3$$
$$= 6.$$

Then, applying Theorem 1.32, we get

$$\int_0^3 f(t) d_g t = f(0)(g(0+) - g(0)) + f(1)(g(1+) - g(1-))$$
$$+ f(2)(g(2+) - g(2-))$$
$$+ f(3)(g(3) - g(3-))$$

$$= 2(0-(-2))+6(1-0)$$
$$= 4+6$$
$$= 10.$$

Exercise 1.6. Let $f(t) = 2t+1$ and
$$g(t) = \begin{cases} 1 & \text{for } t=0, \\ -1 & \text{for } t \in (0,1), \\ 2 & \text{for } t \in (1,2), \\ -3 & \text{for } t \in (2,3), \\ 0 & \text{for } t \in (3,4), \\ 1 & \text{for } t \in (4,5), \\ -1 & \text{for } t=5. \end{cases}$$

Using Theorem 1.32, compute the integral
$$\int_0^5 f(t) d_g t.$$

Answer 1.3. -10.

1.7 Passage to the Limit Under the Stieltjes Integral Sign

Theorem 1.33. *Let $f \in \mathscr{C}([a,b])$ and g be of bounded variation on $[a,b]$. Then,*
$$\left| \int_a^b f(t) d_g t \right| \le \max_{t \in [a,b]} \bigvee_a^b (g).$$

Proof. Let P be a partition of $[a,b]$ given by (1.13) and $\xi_k \in [t_k, t_{k+1}]$, $k \in \{0, \ldots, n-1\}$. Then, we have
$$|\sigma(P,f,g)| = \left| \sum_{j=0}^{n-1} f(\xi_j)(g(t_{k+1}) - g(t_k)) \right|$$
$$\le \sum_{j=0}^{n-1} |f(\xi_j)||g(t_{k+1}) - g(t_k)|$$

$$\leq \max_{t\in[a,b]} |f(t)| \sum_{j=0}^{n-1} |g(t_{j+1}) - g(t_j)|$$

$$\leq \max_{t\in[a,b]} |f(t)| \bigvee_a^b (g).$$

Hence,

$$\left| \int_a^b f(t)d_g t \right| = \left| \lim_{\lambda(P)\to 0} \sigma(P,f,g) \right|$$

$$\leq \max_{t\in[a,b]} |f(t)| \bigvee_a^b (g).$$

This completes the proof. □

Theorem 1.34. *Let $\{f_n\}_{n\in\mathbb{N}}$ be a sequence of continuous functions on $[a,b]$ that converges uniformly to the continuous function f on $[a,b]$. Also, let g be a function of bounded variation on $[a,b]$. Then,*

$$\lim_{n\to\infty} \int_a^b f_n(t)d_g t = \int_a^b f(t)d_g t.$$

Proof. Applying Theorem 1.33, we get

$$\left| \int_a^b f_n(t)d_g t - \int_a^b f(t)d_g t \right| = \left| \int_a^b (f_n(t) - f(t))d_g t \right|$$

$$\leq \max_{t\in[a,b]} |f_n(t) - f(t)| \bigvee_a^b (g)$$

$$\to 0 \quad \text{as } n \to \infty.$$

This completes the proof. □

Theorem 1.35 (The Helly Second Theorem). *Let $f \in \mathscr{C}([a,b])$ and $\{g_n\}_{n\in\mathbb{N}}$ be a sequence of functions finite and defined on $[a,b]$*

that converges to a finite function g at every point of $[a,b]$. If
$$\bigvee_a^b (g_n) < K, \quad n \in \mathbb{N},$$
for some positive constant K, then
$$\lim_{n \to \infty} \int_a^b f(t) d_{g_n} t = \int_a^b f(t) d_g t.$$

Proof. Consider the partition P of $[a,b]$ given by (1.13). Then,
$$\sum_{j=0}^{k-1} |g_m(t_{j+1}) - g_m(t_j)| < K, \quad m \in \mathbb{N},$$
whereupon
$$\lim_{m \to \infty} \sum_{j=0}^{k-1} |g_m(t_{j+1}) - g_m(t_j)| \leq K,$$
i.e.,
$$\sum_{j=0}^{n-1} |g(t_{j+1}) - g(t_j)| \leq K$$
and
$$\bigvee_a^b (g) \leq K.$$

Take $\varepsilon > 0$ arbitrarily. Let the partition P be such that
$$\max_{t \in [t_k, t_{k+1}]} f(t) - \min_{t \in [x_k, t_{k+1}]} f(t) < \frac{\varepsilon}{3K}, \quad k \in \{0, \ldots, n-1\}.$$

Then,
$$\int_a^b f(t) d_g t = \sum_{k=0}^{n-1} \int_{t_k}^{t_{k+1}} f(t) d_g t$$
$$= \sum_{k=0}^{n-1} \int_{t_k}^{t_{k+1}} (f(t) - f(t_k)) d_g t + \sum_{k=0}^{n-1} f(t_k) \int_{t_k}^{t_{k+1}} d_g t$$
$$= \sum_{k=0}^{n-1} \int_{t_k}^{t_{k+1}} (f(t) - f(t_k)) d_g t + \sum_{k=0}^{n-1} f(t_k)(g(t_{k+1}) - g(t_k)).$$

Note that

$$\left|\sum_{k=0}^{n-1}\int_{t_k}^{t_{k+1}}(f(t)-f(t_k))d_g t\right| \le \sum_{k=0}^{n-1}\left|\int_{t_k}^{t_{k+1}}(f(t)-f(t_k))d_g t\right|$$

$$\le \sum_{k=0}^{n-1}\frac{\varepsilon}{3K}\bigvee_{t_k}^{t_{k+1}}(g)$$

$$= \frac{\varepsilon}{3K}\bigvee_{a}^{b}(g)$$

$$< \frac{\varepsilon}{3}.$$

Therefore

$$\int_a^b f(t)d_g t = \sum_{k=0}^{n-1} f(t_k)(g(t_{k+1})-g(t_k)) + \theta\frac{\varepsilon}{3},$$

where $\theta \in [-1,1]$. As above,

$$\int_a^b f(t)d_{g_m} t = \sum_{k=0}^{n-1} f(t_k)(g_m(t_{k+1})-g_m(t_k)) + \theta_m\frac{\varepsilon}{3},$$

where $\theta_m \in [-1,1]$, $m \in \mathbb{N}$. Since

$$\lim_{m\to\infty} g_m(t) = g(t), \quad t \in [a,b],$$

there is an $N \in \mathbb{N}$ so that

$$\left|\sum_{k=0}^{n-1} f(t_k)(g_m(t_{k+1})-g_m(t_k)) - \sum_{k=0}^{n-1} f(t_k)(g(t_{k+1})-g(t_k))\right| < \frac{\epsilon}{3}$$

for any $m > N$, $m \in \mathbb{N}$. Hence,

$$\left|\int_a^b f(t)d_{g_m}t - \int_a^b f(t)d_g t\right| < \varepsilon$$

for any $m > N$, $m \in \mathbb{N}$. This completes the proof. $\square$

1.8 Advanced Practical Problems

Problem 1.1. Find
$$\bigvee_a^b(f),$$
where

(1) $f(t) = t^2$, $a = 0$, $b = 1$,
(2) $f(t) = t^4 + t$, $a = 0$, $b = 2$,
(3) $f(t) = \cos t$, $a = 0$, $b = \frac{\pi}{2}$,
(4) $f(t) = \sin t$, $a = 0$, $b = \frac{\pi}{2}$,
(5) $f(t) = \frac{1}{t}$, $a = 1$, $b = 2$.

Answer 1.4.

(1) 1,
(2) 18,
(3) 1,
(4) 1,
(5) $\frac{1}{2}$.

Problem 1.2. Let f be of bounded variation and differentiable on $[a, b]$ such that f' is integrable in the Riemann sense on $[a, b]$. Then prove that
$$\bigvee_a^b(f) = \int_a^b |f'(t)|dt.$$

Problem 1.3. Using Problem 1.2, solve Problem 1.1.

Problem 1.4. Let $E \subset [0, 1]$ and χ be its characteristic function. Then prove that χ is of finite variation if and only if E is a finite set.

Problem 1.5. Let
$$f(t) = t + 1$$
and
$$g(t) = t^3, \quad t \in [0, 1].$$

Then, using the definition for the Stieltjes integral, compute
$$\int_0^1 f(t)d_g t.$$

Answer 1.5. $\frac{7}{4}$.

Problem 1.6. Let
$$f(t) = t^4 - t$$
and
$$g(t) = t^3, \quad t \in [0, 2].$$
Then, using Theorem 1.31, compute the integral
$$\int_0^2 f(t)d_g t.$$

Answer 1.6. $\frac{300}{7}$.

Problem 1.7. Let $f(t) = 3t - 2$, $t \in [0, 2]$, and
$$g(t) = \begin{cases} 1 & \text{for } t = 0, \\ -1 & \text{for } t \in \left(0, \frac{1}{2}\right), \\ -2 & \text{for } t \in \left(\frac{1}{2}, 1\right), \\ -3 & \text{for } t \in (1, 2), \\ 4 & \text{for } t = 2. \end{cases}$$
Then, using Theorem 1.32, compute the integral
$$\int_0^2 f(t)d_g t.$$

Answer 1.7. $\frac{61}{2}$.

Problem 1.8. Let $f \in \mathscr{C}([a, b])$ and g be finite and defined on $[a, b]$ such that g' is integrable in the Riemann sense on $[a, b]$. Then prove that
$$\left| \int_a^b f(t)d_g t \right| \leq \max_{t \in [a,b]} |f(t)| \int_a^b |g'(t)|dt.$$

Problem 1.9. Let $f \in \mathscr{C}([a,b])$ and g be monotone on $[a,b]$. Then prove that

$$\left|\int_a^b f(t)d_g t\right| \leq \max_{t\in[a,b]} |f(t)||g(b)-g(a)|.$$

Problem 1.10. Let $f \in \mathscr{C}([a,b])$ and g be finite and defined on $[a,b]$ such that g' is summable in the Riemann sense on $[a,b]$ except the points $a < c_1 < \cdots < c_l < b$. Then prove that

$$\int_a^b f(t)d_g t = \int_a^b f(t)g'(t)dt$$
$$+ f(a)(g(a+)-g(a)) + f(b)(g(b)-g(b-))$$
$$+ \sum_{m=1}^{l} f(c_m)(g(c_m+)-g(c_m-)).$$

Chapter 2

The Stieltjes Derivative

In this chapter, we introduce the Stieltjes derivative and explore some of its properties. Two versions of the chain rule are deduced. The Fermat, Rolle, Cauchy and Lagrange types of mean value theorems are formulated and proved. The Jump and continuous components are defined. The primitive function is introduced, and the Leibniz–Newton formula is deduced. The Taylor polynomials are introduced, and the Taylor formula is explored.

2.1 Definition and Examples

Let $I \subseteq \mathbb{R}$. Here and henceforth, we assume that $g \colon I \to \mathbb{R}$ is a monotone nondecreasing nonconstant function which is continuous from the left everywhere.

Definition 2.1. Let $f \colon I \to \mathbb{R}$ be a given function. The derivative with respect to g or g-derivative or the Stieltjes derivative of f at point $x \in I$ is denoted and defined as follows:

(1)
$$f'_g(t) = \lim_{s \to t} \frac{f(s) - f(t)}{g(s) - g(t)} \qquad (2.1)$$

provided g is continuous at t, or

(2)
$$f'_g(t) = \lim_{s \to t+} \frac{f(s) - f(t)}{g(s) - g(t)} \qquad (2.2)$$

provided g is discontinuous at t.

We state that f is g-differentiable or Stieltjes differentiable at $t_0 \in I$ provided $f'_g(t_0)$ exists.

Remark 2.1. The limit (2.1) does not make any sense at the points of the set

$$C_g = \{t \in I : g \text{ is a constant on } (t - \varepsilon, t + \varepsilon) \text{ for some } \varepsilon > 0\}$$

and no matter how f is. This seeming drawback is not important because the set C_g is negligible in a sense to be precise.

Remark 2.2. If $f'(t)$ and $g'(t)$ exist and $g'(t) > 0$, then

$$f'_g(t) = \frac{f'(t)}{g'(t)}.$$

Define the set

$$D_g = \{t \in I : g(t+) - g(t) > 0\},$$

where $g(t+)$ stands for the right-hand limit of g at t. Note that D_g is countable because g is monotone and for each $t \in D_g$, the limit (2.2) exists if and only if $f(t+)$ exists, and, in that case, we have

$$f'_g(t) = \frac{f(t+) - f(t)}{g(t+) - g(t)}.$$

Definition 2.2. For $k \in \mathbb{N}$, define the kth Stieltjes derivative of f at $t \in I$ as

$$f_g^{(k)}(t) = \left(f_g^{(k-1)}\right)'_g(t),$$

where $f_g^0(t) = f(t)$.

Example 2.1. Let $I = \mathbb{R}$ and
$$g(t) = t^3 + t + 1, \quad t \in \mathbb{R},$$
and
$$f(t) = t^4 - t^2 + t - 1, \quad t \in \mathbb{R}.$$
We will find $f'_g(t)$, $f''_g(t)$, $t \in \mathbb{R}$. We have
$$g'(t) = 3t^2 + 1$$
$$> 0, \quad t \in \mathbb{R}.$$
Therefore, g is nondecreasing on $\mathbb{R}$. Moreover, $g \in \mathscr{C}(\mathbb{R})$. Consequently,
$$f'_g(t) = \frac{f'(t)}{g'(t)}$$
$$= \frac{4t^3 - 2t + 1}{3t^2 + 1}, \quad t \in \mathbb{R}.$$
Next,
$$f''_g(t) = \frac{(f'_g)'(t)}{g'(t)}$$
$$= \frac{\frac{(12t^2 - 2)(3t^2 + 1) - 6t(4t^3 - 2t + 1)}{(3t^2+1)^2}}{3t^2 + 1}$$
$$= \frac{36t^4 - 6t^2 + 12t^2 - 2 - 24t^4 + 12t^2 - 6t}{(3t^2 + 1)^3}$$
$$= \frac{12t^4 + 18t^2 - 6t - 2}{(3t^2 + 1)^3}, \quad t \in \mathbb{R}.$$

Example 2.2. Let $I = \mathbb{R}$ and
$$g(t) = \begin{cases} 2 & \text{for } t \in (1, \infty), \\ t + 3 & \text{for } t \in (-\infty, 1], \end{cases}$$
and
$$f(t) = t^2 + t + 1, \quad t \in \mathbb{R}.$$
We will find $f_g^{(k)}(1)$, $k \in \mathbb{N}$. Note that g is nondecreasing on $\mathbb{R}$ and it is continuous from the left everywhere, and it is discontinuous at 1.

Moreover, $g(1+) = 2$ and $g(1) = 4$. Also, observe that $f \in \mathscr{C}(\mathbb{R})$. Then,

$$f'_g(1) = \frac{f(1+) - f(1)}{g(1+) - g(1)}$$
$$= \frac{3 - 3}{2 - 4}$$
$$= 0.$$

Hence,

$$f_g^{(k)}(1) = 0, \quad k \in \mathbb{N}, \ k \geq 2.$$

Example 2.3. Let $I = \mathbb{R}$ and

$$g(t) = \begin{cases} t^2 & \text{for } t \in (0, \infty), \\ 7 & \text{for } t \in (-\infty, 0], \end{cases}$$

and

$$f(t) = \begin{cases} 4 & \text{for } t \in (0, \infty), \\ t + 2 & \text{for } t \in (-\infty, 0]. \end{cases}$$

We will find $f'_g(0)$. Note that g is nondecreasing on $\mathbb{R}$ and it is continuous from the left everywhere, and it is discontinuous at 0. Moreover, we have

$$g(0+) = 0,$$
$$g(0) = 7,$$
$$f(0+) = 4,$$
$$f(0) = 2.$$

Then,

$$f'_g(0) = \frac{f(0+) - f(0)}{g(0+) - g(0)}$$
$$= \frac{4 - 2}{0 - 7}$$
$$= -\frac{2}{7}.$$

Exercise 2.1. Let $I = \mathbb{R}$ and
$$g(t) = t^3 + 2t, \quad t \in \mathbb{R}.$$
Find $f'_g(t)$, $t \in \mathbb{R}$, where

(1)
$$f(t) = \frac{t^2 + t + 1}{t^2 - t + 1}, \quad t \in \mathbb{R},$$

(2)
$$f(t) = t^4 - 3t^2 + 5t - 1, \quad t \in \mathbb{R},$$

(3)
$$f(t) = \sqrt{t^2 + 1}, \quad t \in \mathbb{R}.$$

Answer 2.1.

(1)
$$\frac{2(1 - t^2)}{(t^2 - t + 1)^2 (3t^2 + 2)}, \quad t \in \mathbb{R},$$

(2)
$$\frac{4t^3 - 6t + 5}{3t^2 + 2}, \quad t \in \mathbb{R},$$

(3)
$$\frac{t}{\sqrt{t^2 + 1}(3t^2 + 2)}, \quad t \in \mathbb{R}.$$

Exercise 2.2. Let $I = \mathbb{R}$ and
$$g(t) = 1 + t, \quad t \in I,$$
and
$$f(t) = t^4 - 2t^3 + 3t^2 - t - 2, \quad t \in I.$$
Find $f_g^{(4)}(t)$, $t \in I$.

Answer 2.2. 24.

2.2 Properties of the Stieltjes Derivative

In this section, we explore some of the properties of the Stieltjes derivative. We start with the following useful results.

Theorem 2.1. *Let $t \in I \setminus D_g$ be such that $f'_g(t)$ exists. Then $t \notin C_g$ and f is continuous from the right at t provided that $g(s) > g(t)$ for $s > t$.*

Proof. If $t \in C_g$, then $f'_g(t)$ does not exist. This is a contradiction. Therefore, $t \notin C_g$. Since $t \in I \setminus D_g$ and $t \notin C_g$, and $f'_g(t)$ exists, and $g(s) > g(t)$ for $s > t$, we get

$$f(t+) = f(t),$$

i.e., f is continuous from the left at t. This completes the proof. □

Theorem 2.2. *Let $t \in I \setminus D_g$ be such that $f'_g(t)$ exists. Then, $t \notin C_g$ and f is continuous from the left at t, provided that $g(s) < g(t)$ for $s < t$.*

Proof. If $t \in C_g$, then $f'_g(t)$ does not exist, which is a contradiction. Thus, $t \notin C_g$. Since $t \in I \setminus D_g$ and $t \notin C_g$, and $f'_g(t)$ exists, and $g(s) < g(t)$ for any $s < t$, we obtain

$$f(t-) = f(t),$$

i.e., f is continuous from the left at t. This completes the proof. □

Theorem 2.3. *For $t \in D_g$, $f'_g(t)$ exists if and only if $f(t+)$ exists.*

Proof. Since $t \in D_g$, we have that

$$g(t+) > g(t).$$

Hence, using the definition for the Stieltjes derivative, we conclude that $f'_g(t)$ exists if and only if there exists $f(t+)$. This completes the proof. □

Theorem 2.4. *Let f_1 and f_2 be real-valued functions defined on a neighbourhood of a point $t \in I$. Also, let $g(s) \neq g(t)$ for $s \neq t$. If f_1 and f_2 are Stieltjes differentiable at t, then for $c_1, c_2 \in \mathbb{R}$, we have*

$$(c_1 f_1 + c_2 f_2)'_g(t) = c_1 f'_{1g}(t) + c_2 f'_{2g}(t).$$

Proof. By definition, we have

$$(c_1 f_1 + c_2 f_2)'_g(t)$$

$$= \begin{cases} \lim_{s \to t} \frac{(c_1 f_1 + c_2 f_2)(s) - (c_1 f_1 + c_2 f_2)(t)}{g(s) - g(t)} & \text{if } g \text{ is continuous at } t, \\ \lim_{s \to t+} \frac{(c_1 f_1 + c_2 f_2)(s) - (c_1 f_1 + c_2 f_2)(t)}{g(s) - g(t)} & \text{if } g \text{ is discontinuous at } t \end{cases}$$

$$= \begin{cases} \lim_{s \to t} \frac{c_1(f_1(s) - f_1(t)) + c_2(f_2(s) - f_2(t))}{g(s) - g(t)} & \text{if } g \text{ is continuous at } t, \\ \lim_{s \to t+} \frac{c_1(f_1(s) - f_1(t)) + c_2(f_2(s) - f_2(t))}{g(s) - g(t)} & \text{if } g \text{ is discontinuous at } t \end{cases}$$

$$= c_1 \begin{cases} \lim_{s \to t} \frac{f_1(s) - f_1(t)}{g(s) - g(t)} & \text{if } g \text{ is continuous at } t, \\ \lim_{s \to t+} \frac{f_1(s) - f_1(t)}{g(s) - g(t)} & \text{if } g \text{ is discontinuous at } t \end{cases}$$

$$+ c_2 \begin{cases} \lim_{s \to t} \frac{f_2(s) - f_2(t)}{g(s) - g(t)} & \text{if } g \text{ is continuous at } t, \\ \lim_{s \to t+} \frac{f_2(s) - f_2(t)}{g(s) - g(t)} & \text{if } g \text{ is discontinuous at } t \end{cases}$$

$$= c_1 f'_{1g}(t) + c_2 f'_{2g}(t).$$

This completes the proof. $\square$

Theorem 2.5. *Let f_1 and f_2 be real-valued functions defined on a neighbourhood of a point $t \in I$. Also, let $g(s) \neq g(t)$ for $s \neq t$. If f_1 and f_2 are Stieltjes differentiable at t, then*

$$(f_1 f_2)'_g(t) = f'_{1g}(t) f_2(t+) + f_1(t+) f'_{2g}(t).$$

Proof. By definition, we have

$$(f_1 f_2)'_g(t)$$

$$= \begin{cases} \lim_{s \to t} \frac{(f_1 f_2)(s) - (f_1 f_2)(t)}{g(s) - g(t)} & \text{if } g \text{ is continuous at } t, \\ \lim_{s \to t+} \frac{(f_1 f_2)(s) - (f_1 f_2)(t)}{g(s) - g(t)} & \text{if } g \text{ is discontinuous at } t \end{cases}$$

$$= \begin{cases} \lim_{s \to t} \frac{f_1(s) f_2(s) - f_1(t) f_2(t)}{g(s) - g(t)} & \text{if } g \text{ is continuous at } t, \\ \lim_{s \to t+} \frac{f_1(s) f_2(s) - f_1(t) f_2(t)}{g(s) - g(t)} & \text{if } g \text{ is discontinuous at } t \end{cases}$$

$$= \begin{cases} \lim\limits_{s \to t} \dfrac{(f_1(s)-f_1(t))f_2(s)+f_1(t)(f_2(s)-f_2(t))}{g(s)-g(t)} \\ \quad \text{if } g \text{ is continuous at } t, \\ \lim\limits_{s \to t+} \dfrac{(f_1(s)-f_1(t))f_2(s)+f_1(t)(f_2(s)-f_2(t))}{g(s)-g(t)} \\ \quad \text{if } g \text{ is discontinuous at } t \end{cases}$$

$$= f_2(t+) \begin{cases} \lim\limits_{s \to t} \dfrac{f_1(s)-f_1(t)}{g(s)-g(t)} & \text{if } g \text{ is continuous at } t, \\ \lim\limits_{s \to t+} \dfrac{f_1(s)-f_1(t)}{g(s)-g(t)} & \text{if } g \text{ is discontinuous at } t \end{cases}$$

$$+ f_1(t+) \begin{cases} \lim\limits_{s \to t} \dfrac{f_2(s)-f_2(t)}{g(s)-g(t)} & \text{if } g \text{ is continuous at } t, \\ \lim\limits_{s \to t+} \dfrac{f_2(s)-f_2(t)}{g(s)-g(t)} & \text{if } g \text{ is discontinuous at } t \end{cases}$$

$$= f_2(t+)f'_{1g}(t) + f_1(t+)f'_{2g}(t).$$

This completes the proof. □

Theorem 2.6. *Let f_1 and f_2 be real-valued functions defined on a neighbourhood of a point $t \in I$. Also, let $g(s) \neq g(t)$ for $s \neq t$. If f_1 and f_2 are Stieltjes differentiable at t and $f_2(t)f_2(t+) \neq 0$, then*

$$\left(\dfrac{f_1}{f_2}\right)'_g(t) = \dfrac{f'_{1g}(t)f_2(t) - f_1(t)f'_{2g}(t)}{f_2(t)f_2(t+)}.$$

Proof. By definition, we have

$$\left(\dfrac{f_1}{f_2}\right)'_g(t) = \begin{cases} \lim\limits_{s \to t} \dfrac{\left(\frac{f_1}{f_2}\right)(s)-\left(\frac{f_1}{f_2}\right)(t)}{g(s)-g(t)} & \text{if } g \text{ is continuous at } t, \\ \lim\limits_{s \to t+} \dfrac{\left(\frac{f_1}{f_2}\right)(s)-\left(\frac{f_1}{f_2}\right)(t)}{g(s)-g(t)} & \text{if } g \text{ is discontinuous at } t \end{cases}$$

$$= \begin{cases} \lim\limits_{s \to t} \dfrac{\frac{f_1(s)}{f_2(s)}-\frac{f_1(t)}{f_2(t)}}{g(s)-g(t)} & \text{if } g \text{ is continuous at } t, \\ \lim\limits_{s \to t+} \dfrac{\frac{f_1(s)}{f_2(s)}-\frac{f_1(t)}{f_2(t)}}{g(s)-g(t)} & \text{if } g \text{ is discontinuous at } t \end{cases}$$

$$= \begin{cases} \lim\limits_{s \to t} \dfrac{f_1(s)f_2(t)-f_1(t)f_2(s)}{f_2(s)f_2(t)(g(s)-g(t))} & \text{if } g \text{ is continuous at } t, \\ \lim\limits_{s \to t+} \dfrac{f_1(s)f_2(t)-f_1(t)f_2(s)}{f_2(s)f_2(t)(g(s)-g(t))} & \text{if } g \text{ is discontinuous at } t \end{cases}$$

$$= \begin{cases} \lim_{s \to t} \frac{(f_1(s)-f_1(t))f_2(t)-f_1(t)(f_2(s)-f_2(t))}{f_2(s)f_2(t)(g(s)-g(t))} \\ \quad \text{if } g \text{ is continuous at } t, \\ \lim_{s \to t+} \frac{(f_1(s)-f_1(t))f_2(t)-f_1(t)(f_2(s)-f_2(t))}{f_2(s)f_2(t)(g(s)-g(t))} \\ \quad \text{if } g \text{ is discontinuous at } t \end{cases}$$

$$= \frac{1}{f_2(t+)} \begin{cases} \lim_{s \to t} \frac{f_1(s)-f_1(t)}{g(s)-g(t)} & \text{if } g \text{ is continuous at } t, \\ \lim_{s \to t+} \frac{f_1(s)-f_2(t)}{g(s)-g(t)} & \text{if } g \text{ is discontinuous at } t \end{cases}$$

$$- \frac{f_1(t)}{f_2(t)f_2(t+)} \begin{cases} \lim_{s \to t} \frac{f_2(s)-f_2(t)}{g(s)-g(t)} & \text{if } g \text{ is continuous at } t, \\ \lim_{s \to t+} \frac{f_2(s)-f_2(t)}{g(s)-g(t)} & \text{if } g \text{ is discontinuous at } t \end{cases}$$

$$= \frac{1}{f_2(t+)} f'_{1g}(t) - \frac{f_1(t)}{f_2(t)f_2(t+)} f'_{2g}(t)$$

$$= \frac{f_2(t)f'_{1g}(t) - f_1(t)f'_{2g}(t)}{f_2(t)f_2(t+)}.$$

This completes the proof. □

Example 2.4. Let $I = \mathbb{R}$ and

$$g(t) = 3t + \sin t, \quad t \in \mathbb{R},$$

and

$$f(t) = (t^2 + t + 1)(t^2 - t + 1), \quad t \in \mathbb{R}.$$

We will find $f'_g(t)$, $t \in \mathbb{R}$. Note that

$$g'(t) = 3 + \cos t$$
$$> 0, \quad t \in \mathbb{R}.$$

Therefore, g is an increasing function on $\mathbb{R}$. Moreover, $g \in \mathscr{C}(\mathbb{R})$. Set $f(t) = f_1(t)f_2(t)$, $t \in \mathbb{R}$, where

$$f_1(t) = t^2 + t + 1, \quad t \in \mathbb{R},$$

and
$$f_2(t) = t^2 - t + 1, \quad t \in \mathbb{R}.$$
Then,
$$f_1'(t) = 2t + 1$$
and
$$f_2'(t) = 2t - 1.$$
Hence,
$$f_{1g}'(t) = \frac{f_1'(t)}{g'(t)}$$
$$= \frac{2t+1}{3+\cos t}$$
and
$$f_{2g}'(t) = \frac{f_2'(t)}{g'(t)}$$
$$= \frac{2t-1}{3+\cos t}.$$
Now, applying Theorem 2.5, we arrive at
$$f_g'(t) = f_{1g}'(t)f_2(t) + f_1(t)f_{2g}'(t)$$
$$= \frac{2t+1}{3+\cos t}(t^2 - t + 1) + (t^2 + t + 1)\frac{2t-1}{3+\cos t}$$
$$= \frac{(2t+1)(t^2 - t + 1) + (t^2 + t + 1)(2t-1)}{3+\cos t}$$
$$= \frac{2t^3 - 2t^2 + 2t + t^2 - t + 1 + 2t^3 - t^2 + 2t^2 - t + 2t - 1}{3+\cos t}$$
$$= \frac{4t^3 + 2t}{3+\cos t}.$$
Thus,
$$f_g'(t) = \frac{4t^3 + 2t}{3+\cos t}, \quad t \in \mathbb{R}.$$

Example 2.5. Let g be as in Example 2.4 and

$$f(t) = \frac{t^2 + t + 1}{t^2 - t + 1}, \quad t \in \mathbb{R}.$$

We will find $f'_g(t)$, $t \in \mathbb{R}$. We write

$$f(t) = \frac{f_1(t)}{f_2(t)}, \quad t \in \mathbb{R},$$

where f_1 and f_2 are as in Example 2.4. By the computations in Example 2.4 and applying Theorem 2.6, we find

$$\begin{aligned} f'_g(t) &= \frac{f'_{1g}(t) f_2(t) - f_1(t) f - 2g'(t)}{(f_2(t))^2} \\ &= \frac{\frac{2t+1}{3+\cos t}(t^2 - t + 1) - (t^2 + t + 1)\frac{2t-1}{3+\cos t}}{(t^2 - t + 1)^2} \\ &= \frac{(2t+1)(t^2 - t + 1) - (t^2 + t + 1)(2t - 1)}{(3 + \cos t)(t^2 - t + 1)^2} \\ &= \frac{2t^3 - 2t^2 + 2t + t^2 - t + 1 - 2t^3 + t^2 - 2t^2 + t - 2t + 1}{(3 + \cos t)(t^2 - t + 1)^2} \\ &= \frac{2(-t^2 + 1)}{(3 + \cos t)(t^2 - t + 1)^2}. \end{aligned}$$

Thus,

$$f'_g(t) = \frac{2(-t^2 + 1)}{(3 + \cos t)(t^2 - t + 1)^2}, \quad t \in \mathbb{R}.$$

Exercise 2.3. Let $I = \mathbb{R}$ and

$$\begin{aligned} g(t) &= 1 + 3t^3, \\ f_1(t) &= t^4 + t^2 + 1, \quad \text{and} \\ f_2(t) &= t^2 + 1, \quad t \in \mathbb{R}. \end{aligned}$$

Check Theorems 2.5 and 2.6.

2.3 Chain Rules

We have two versions of the chain rule for the Stieltjes derivatives of composition of functions. Note that, in both versions, the usual derivative is involved.

Theorem 2.7 (**The Chain Rule I**). *Let f be a real-valued function defined in a neighbourhood of $t \in I \setminus D_g$ and let h be a real-valued function defined in a neighbourhood of $f(t)$. If $h'(f(t))$ and $f'_g(t)$ both exist, then*

$$(h \circ f)'_g(t) = h'(f(t))f'_g(t) \quad for\ t \in I \setminus D_g.$$

Proof. By definition, we have

$(h \circ f)'_g(t)$

$$= \begin{cases} \lim_{s \to t} \frac{h(f(s)) - h(f(t))}{g(s) - g(t)} & \text{if } g \text{ is continuous at } t, \\ \lim_{s \to t+} \frac{h(f(s)) - h(f(t))}{g(s) - g(t)} & \text{if } g \text{ is discontinuous at } t \end{cases}$$

$$= \begin{cases} \lim_{s \to t} \left(\frac{h(f(s)) - h(f(t))}{f(s) - f(t)} \right) \left(\frac{f(s) - f(t)}{g(s) - g(t)} \right) & \text{if } g \text{ is continuous at } t, \\ \lim_{y \to t+} \left(\frac{h(f(s)) - h(f(t))}{f(s) - f(t)} \right) \left(\frac{f(s) - f(t)}{g(s) - g(t)} \right) & \text{if } g \text{ is discontinuous at } t \end{cases}$$

$$= h'(f(t))f'_g(t).$$

Thus,

$$(h \circ f)'_g(t) = h'(f(t))f'_g(t) \quad for\ t \in I \setminus D_g.$$

This completes the proof. $\square$

Theorem 2.8 (**The Chain Rule II**). *Let f be a real-valued function defined in a neighbourhood of $t \in I \setminus D_g$ and let h be a real-valued function defined in a neighbourhood of $f(t)$. If $h'_g(f(t))$, $g'(f(t))$, and $f'_g(t)$ exist, then*

$$(h \circ f)'_g(t) = h'_g(f(t))g'(f(t))f'_g(t) \quad for\ t \in I \setminus D_g.$$

Proof. By definition, we have

$(h \circ f)'_g(t)$

$$= \begin{cases} \lim_{s \to t} \frac{h(f(s))-h(f(t))}{g(s)-g(t)} & \text{if } g \text{ is continuous at } t, \\ \lim_{s \to t+} \frac{h(f(s))-h(f(t))}{g(s)-g(t)} & \text{if } g \text{ is discontinuous at } t \end{cases}$$

$$= \begin{cases} \lim_{s \to t} \left(\frac{h(f(s))-h(f(t))}{g(f(s))-g(f(t))}\right) \left(\frac{g(f(s))-g(f(t))}{f(s)-f(t)}\right) \left(\frac{f(s)-f(t)}{g(s)-g(t)}\right) \\ \quad \text{if } g \text{ is continuous at } t, \\ \lim_{s \to t+} \left(\frac{h(f(s))-h(f(t))}{g(f(s))-g(f(t))}\right) \left(\frac{g(f(s))-g(f(t))}{f(s)-f(t)}\right) \left(\frac{f(s)-f(t)}{g(s)-g(t)}\right) \\ \quad \text{if } g \text{ is discontinuous at } t \end{cases}$$

$$= h'_g(f(t))g'(f(t))f'_g(t).$$

Thus,

$$(h \circ f)'_g(t) = h'_g(f(t))g'(f(t))f'_g(t) \quad \text{for } t \in I \setminus D_g.$$

This completes the proof. □

2.4 Mean Value Theorems

In this section, we explore some mean value theorems. We start with the Fermat theorem.

Theorem 2.9 (The Fermat Theorem). *Let $f, g \colon I \to \mathbb{R}$ be continuous functions at $t_0 \in I$ and t_0 is a point of extremum of f. If f is Stieltjes differentiable at t_0, then $f'_g(t_0) = 0$.*

Proof. Without any restriction, suppose that t_0 is a point of minimum of f. Then, there is a neighbourhood V of t_0 so that $f(t) \geq f(t_0)$ for $t \in V$. Since g is nondecreasing, we have

$$g(t_0+) \geq g(t_0)$$

and

$$g(t_0-) \leq g(t_0).$$

Then
$$\frac{f(t_0-) - f(t_0)}{g(t_0-) - g(t_0)} \leq 0,$$
whereupon
$$f'_g(t_0) \leq 0 \qquad (2.3)$$
and
$$\frac{f(t_0+) - f(t_0)}{g(t_0+) - g(t_0)} \geq 0,$$
from where
$$f'_g(t_0) \geq 0. \qquad (2.4)$$
Now, in view of (2.3) and (2.4), we get $f'_g(t_0) = 0$. This completes the proof. $\square$

Example 2.6. Let $I = \mathbb{R}$ and
$$f(t) = t^3 - 12t, \quad t \in \mathbb{R},$$
and
$$g(t) = \begin{cases} 1 & \text{for } t \in (1, \infty), \\ t+4 & \text{for } t \in (-\infty, 1]. \end{cases}$$
Then we have that
$$f'(t) = 3t^2 - 12$$
$$= 3(t^2 - 4)$$
$$= 3(t-2)(t+2), \quad t \in \mathbb{R}.$$
Hence,
$$f'(t) = 0, \quad t \in \mathbb{R},$$
if and only if
$$t = -2 \quad \text{or} \quad t = 2.$$
Next,
$$f''(t) = 6t, \quad t \in \mathbb{R}.$$
Therefore, f has a local maximum at the point $t = 2$ and a local minimum at $t = -2$. Note that f is not Stieltjes differentiable at

$t = 2$, and it is Stieltjes differentiable at $t = -2$. By the Fermat Theorem (Theorem 2.9), it follows that

$$f'_g(-2) = 0.$$

Actually,

$$f'_g(-2) = \frac{f'(-2)}{g'(-2)}$$
$$= 0.$$

Theorem 2.10 (**The Rolle Theorem**). *Let $I = [a,b]$ and $f \in \mathscr{C}([a,b])$ be Stieltjes differentiable on (a,b) and $f(a) = f(b)$. Then, there exists $c \in (a,b)$ such that*

$$f'_g(c) = 0. \qquad (2.5)$$

Proof. Suppose that $t_0 \in D_g$. Then, $f'_g(t_0)$ exists if and only if $f(t_0+)$ exists. Since $f \in \mathscr{C}([a,b])$, we have that $f(t_0+) = f(t_0)$, and then

$$f'_g(t_0) = 0.$$

If f is a constant function on $[a,b]$, then, for $t \in [a,b]$, we have

$$f'_g(t) = 0.$$

Assume that f is not a constant function on $[a,b]$. Since $f \in \mathscr{C}([a,b])$ and $[a,b]$ is a compact set, there is a point $c \in (a,b)$ that is an extremum point of f. Now, applying the Fermat theorem (Theorem 2.9), we get (2.5). This completes the proof. □

Example 2.7. Let f and g be as in Example 2.6. Also, let $a = -2\sqrt{3}$ and $b = 0$. Then

$$f(t) = t(t^2 - 12)$$
$$= t(t - 2\sqrt{3})(t + 2\sqrt{3}), \quad t \in \mathbb{R},$$

and we have that

$$f(a) = f(b) = 0.$$

The function f is Stieltjes differentiable on $[a,b]$ and
$$f'_g(t) = \frac{f'(t)}{g'(t)}$$
$$= 3(t-2)(t+2), \quad t \in [a,b].$$

Hence,
$$f'_g(t) = 0, \quad t \in [a,b],$$
if and only if $t = -2$. Thus, $c = -2$ in the Rolle Theorem.

Theorem 2.11 (**The Cauchy Mean Value Theorem**). *Let $I = [a,b]$ and $f, h \in \mathscr{C}([a,b])$ be Stieltjes differentiable on (a,b) and $h'_g(t) \neq 0$ for $t \in (a,b)$. Then, $h(a) \neq h(b)$ and there exists $c \in (a,b)$ such that*
$$\frac{f(b) - f(a)}{h(b) - h(a)} = \frac{f'_g(c)}{h'_g(c)}. \tag{2.6}$$

Proof. Suppose that $h(a) = h(b)$. Then, by Theorem 2.10, it follows that there is $c \in (a,b)$, for which $h'_g(c) = 0$. This is a contradiction. Therefore, $h(a) \neq h(b)$. Define the function $\phi \colon [a,b] \to \mathbb{R}$ as
$$\phi(t) = f(t) - \frac{f(b) - f(a)}{h(b) - h(a)} h(t), \quad t \in [a,b].$$

Then, $\phi \in \mathscr{C}([a,b])$ and ϕ is Stieltjes differentiable on (a,b) with
$$\phi'_g(t) = f'_g(t) - \frac{f(b) - f(a)}{h(b) - h(a)} h'_g(c). \tag{2.7}$$

Moreover,
$$\phi(b) - \phi(a) = f(b) - \frac{f(b) - f(a)}{h(b) - h(a)} h(b) - f(a) + \frac{f(b) - f(a)}{h(b) - h(a)} h(a)$$
$$= f(b) - f(a) - \frac{f(b) - f(a)}{h(b) - h(a)} (h(b) - h(a))$$
$$= 0,$$

i.e.,
$$\phi(b) = \phi(a).$$

Now, applying the Rolle theorem (Theorem 2.10), it follows that there exists $c \in (a,b)$ such that
$$\phi'_g(c) = 0.$$
This together with (2.7) yields
$$f'_g(c) - \frac{f(b) - f(a)}{h(b) - h(a)} h'_g(c) = 0,$$
whereupon we get the desired result. This completes the proof. □

Example 2.8. Let f and g be as in Example 2.6 and
$$h(t) = t^2 + 1, \quad t \in \mathbb{R},$$
and $a = -2\sqrt{3}$, $b = -1$. Then
$$f(b) - f(a) = (-1)^3 - 12 \cdot (-1)$$
$$= -1 + 12$$
$$= 11,$$
$$h(b) - h(a) = (-1)^2 + 1 - (-2\sqrt{3})^2 - 1$$
$$= 1 - 12$$
$$= -11,$$
and
$$h'_g(t) = \frac{h'(t)}{g'(t)}$$
$$= 2t$$
$$\neq 0, \quad t \in [a,b].$$
Hence,
$$\frac{f'_g(t)}{h'_g(t)} = \frac{f(b) - f(a)}{h(b) - h(a)}, \quad t \in (a,b),$$
if and only if
$$\frac{3(t-2)(t+2)}{2t} = -1, \quad t \in (a,b),$$

if and only if
$$3(t^2 - 4) - 2t, \quad t \in (a,b),$$
if and only if
$$3t^2 + 2t - 12 = 0, \quad t \in (a,b),$$
if and only if
$$t = \frac{-1 - \sqrt{37}}{3}.$$

Thus,
$$c = \frac{-1 - \sqrt{37}}{3}$$
in the Cauchy mean value theorem.

Theorem 2.12 (**The Lagrange Mean Value Theorem**). *Let $I = [a,b]$ and $f, g \in \mathscr{C}([a,b])$ be such that f is Stieltjes differentiable on (a,b). Then, there exists $c \in (a,b)$ such that*
$$f(b) - f(a) = f'_g(c)(g(b) - g(a)).$$

Proof. Applying Theorem 2.11 for $h = g$ and using the fact that $g'_g(t) = 1$, $t \in [a,b]$, we get the desired result. This completes the proof. □

Example 2.9. Let f and g be as in Example 2.6, and $a = -2\sqrt{3}$, $b = -1$. Then,
$$g(b) - g(a) = -1 + 4 + 2\sqrt{3} - 4$$
$$= 2\sqrt{3} - 1.$$

Hence,
$$f'_g(t) = \frac{f(b) - f(a)}{g(b) - g(a)}, \quad t \in (a,b),$$
if and only if
$$3(t-2)(t+2) = \frac{11}{2\sqrt{3} - 1}, \quad t \in (a,b),$$

if and only if
$$3t^2 - 12 = \frac{11(2\sqrt{3}+1)}{11}, \quad t \in (a,b),$$
if and only if
$$3t^2 = 2\sqrt{3} + 13, \quad t \in (a,b),$$
if and only if
$$t^2 = \frac{2\sqrt{3}+13}{3}, \quad t \in (a,b),$$
if and only if
$$t = -\sqrt{\frac{2\sqrt{3}+13}{3}}.$$

Thus,
$$c = -\sqrt{\frac{2\sqrt{3}+13}{3}}$$
in the Lagrange Mean Value Theorem.

Theorem 2.13. *Let $I = [a,b]$ and g has a finite number of points of discontinuity. Define the saltus function $S^g : [a,b] \to \mathbb{R}$ as follows:*
$$S^g(t) = (g(a+) - g(a)) + \sum_{t_k < t}(g(t_k+) - g(t_k-)) + (g(t) - g(t-)),$$
$$t \in [a,b],$$
where t_k are points of discontinuity of g. Then, S^g is Stieltjes differentiable on (a,b).

Proof. For simplicity, suppose that g has only point of discontinuity t_0. Then,
$$S^g(t) = \begin{cases} 0 & \text{for } t \in [a, t_0), \\ g(t_0) - g(t_0-) & \text{for } t = t_0, \\ g(t_0+) - g(t_0-) & \text{for } t \in (t_0, b]. \end{cases}$$

Since S^g is continuous on $(a,b) \setminus \{t_0\}$, it is Stieltjes differentiable on $(a,b) \setminus \{t_0\}$ and
$$S^g g'(t) = 0, \quad t \in (a,b) \setminus \{t_0\}.$$
Next,
$$\begin{aligned} S^g g'(t_0) &= \lim_{s \to t_0+} \frac{S^g(y) - S^g(t_0)}{g(s) - g(t_0)} \\ &= \lim_{s \to t_0+} \frac{g(s) - g(t_0-) - g(t_0) + g(t_0-)}{g(s) - g(t_0)} \\ &= 1. \end{aligned}$$
This completes the proof. $\square$

Remark 2.3. Let $D_{[a,b]}$ denote the set of all increasing functions whose saltus functions are Stieltjes differentiable on (a,b). Let $h \in D_{[a,b]}$ and S^h be its saltus function. Then the function $\overline{h} \colon [a,b] \to \mathbb{R}$ defined by
$$\overline{h}(t) = h(t) - S^h(t), \quad t \in [a,b],$$
is continuous on $[a,b]$. We state that the function $\overline{h}$ is the continuous component of h.

Theorem 2.14. Let $g \in D_{[a,b]}$. Then $\overline{g}$ is Stieltjes differentiable on (a,b).

Proof. By Theorem 2.13, we get that S^g is the Stieltjes differentiable function on (a,b). Applying Theorem 2.4, we obtain that $\overline{g}$ is the Stieltjes differentiable function on (a,b) and
$$\begin{aligned} \overline{g}'_g(t) &= g'_g(t) - S^g g'(t) \\ &= 1 - \begin{cases} 0 & \text{if } g \text{ is continuous at } t, \\ 1 & \text{if } g \text{ is discontinuous at } t \end{cases} \\ &= \begin{cases} 1 & \text{if } g \text{ is continuous at } x, \\ 0 & \text{if } g \text{ is discontinuous at } x. \end{cases} \end{aligned}$$
This completes the proof. $\square$

2.5 Stieltjes Primitive and the Leibniz–Newton–Stieltjes Formula

Definition 2.3. Let $f\colon I \to \mathbb{R}$ be a given function. A function $F\colon I \to \mathbb{R}$ is said to be a Stieltjes primitive of f provided
$$F'_g(t) = f(t), \quad t \in I.$$

Remark 2.4. A Stieltjes primitive of a given function $f\colon I \to \mathbb{R}$ may not be unique.

Theorem 2.15. *Let* $I = [a,b]$ *and* $f \in \mathscr{C}([a,b])$ *and define*
$$F(t) = \int_a^t f(\tau) d_g\tau \quad \text{for } t \in [a,b].$$

Then, F is a Stieltjes primitive of f and
$$F'_g(t) = f(t) \quad \text{for } t \in [a,b].$$

Proof. Applying Theorem 2.13, we get

$$F'_g(t) = \begin{cases} \lim\limits_{s \to t} \frac{F(s)-F(t)}{g(s)-g(t)} & \text{if } g \text{ is continuous at } t, \\ \lim\limits_{s \to t+} \frac{F(s)-F(t)}{g(s)-g(t)} & \text{if } g \text{ is discontinuous at } t \end{cases}$$

$$= \begin{cases} \lim\limits_{s \to t} \frac{\int_a^s f(\tau)d_g\tau - \int_a^t f(\tau)d_g\tau}{g(s)-g(t)} & \text{if } g \text{ is continuous at } t, \\ \lim\limits_{s \to t+} \frac{\int_a^s f(t)d_gt - \int_a^t f(\tau)d_g\tau}{g(s)-g(t)} & \text{if } g \text{ is discontinuous at } t \end{cases}$$

$$= \begin{cases} \lim\limits_{s \to t} \frac{\int_t^s f(\tau)d_g\tau}{g(s)-g(t)} & \text{if } g \text{ is continuous at } t, \\ \lim\limits_{s \to t+} \frac{\int_t^s f(\tau)d_g\tau}{g(s)-g(t)} & \text{if } g \text{ is discontinuous at } t \end{cases}$$

$$= \begin{cases} \lim\limits_{s \to t} \frac{f(\xi)(g(s)-g(t))}{g(s)-g(t)}, \quad t \leq \xi \leq s, & \text{if } g \text{ is continuous at } t, \\ \lim\limits_{s \to t+} \frac{f(\eta)(g(s)-g(t))}{g(s)-g(t)}, \quad t+ \leq \eta \leq s, & \text{if } g \text{ is discontinuous at } t \end{cases}$$

$$= f(t).$$

Thus,
$$F'_g(t) = f(t) \quad \text{for } t \in [a,b].$$

This completes the proof. $\square$

Theorem 2.16 (The Leibniz–Newton–Stieltjes Formula). Let $I = \mathbb{R}$. Assume that $f\colon [a,b] \to \mathbb{R}$ is Stieltjes integrable and $F \in \mathscr{C}([a,b])$ is its Stieltjes primitive. Then,
$$\int_a^b f(\tau) d_g \tau = F(b) - F(a).$$

Proof. Let P be a partition of the interval $[a,b]$ given by
$$a = t_0 < t_1 < \cdots < t_n = b.$$
Applying Theorem 2.12, we find
$$F(b) - F(a) = \sum_{j=1}^n (F(t_j) - F(t_{j-1}))$$
$$= \sum_{j=1}^n F'_g(\xi_j)(g(t_j) - g(t_{j-1}))$$
$$= \sum_{j=1}^n f(\xi_j)(g(t_j) - g(t_{j-1}))$$
$$= \sigma(P, f, g),$$
where $\xi_j \in (t_{j-1}, t_j)$. Since f is Stieltjes integrable on $[a,b]$, we obtain
$$F(b) - F(a) = \lim_{\lambda(P) \to 0} \sigma(P, f, g)$$
$$= \int_a^b f(t) d_g t.$$
This completes the proof. □

Example 2.10. Let f and g be as in Example 2.6. Also, let
$$F(t) = \frac{t^4}{4} - 6t^2, \quad t \in [-3, 0].$$
Then,
$$F'_g(t) = \frac{F'(t)}{g'(t)}$$
$$= t^3 - 12t, \quad t \in [-3, 0].$$
Thus, F is a Stieltjes primitive of f on $[-3, 0]$.

2.6 Stieltjes Monomials and the Stieltjes–Taylor Formula

In this section, we introduce the Stieltjes monomials $h_k \colon I \times I \to \mathbb{R}$, $k \in \mathbb{N}_0$, as follows

$$h_0(t,s) = 1,$$

$$h_k(t,s) = \int_s^t h_{k-1}(u,s) d_g u, \quad k \in \mathbb{N},\ t,s \in I.$$

Note that

$$h'_{kg}(t,s) = h_{k-1}(t,s), \quad k \in \mathbb{N},\ t,s \in I.$$

Example 2.11. Let $I = \mathbb{R}$ and

$$g(t) = t^2 + 1, \quad t \in \mathbb{R}.$$

Then we have the following:

$$h_0(t,s) = 1, \quad t,s \in \mathbb{R},$$

$$h_1(t,s) = \int_s^t h_0(u,s) d_g u$$

$$= \int_s^t d_g u$$

$$= g(t) - g(s)$$

$$= t^2 + 1 - s^2 - 1$$

$$= t^2 - s^2, \quad t,s \in \mathbb{R},$$

and

$$h_2(t,s) = \int_s^t h_1(u,s) d_g u$$

$$= \int_s^t (u^2 - s^2) g'(u) du$$

$$= 2 \int_s^t (u^2 - s^2) u\, du$$

$$= \int_s^t (u^2 - s^2) d(u^2 - s^2)$$

$$= \frac{(u^2 - s^2)^2}{2} \Big|_{u=s}^{u=t}$$

$$= \frac{(t^2 - s^2)^2}{2}, \quad t, s \in \mathbb{R}.$$

Assume that

$$h_k(t, s) = \frac{(t^2 - s^2)^k}{k!}, \quad t, s \in \mathbb{R},$$

for some $k \in \mathbb{N}$. We will prove that

$$h_{k+1}(t, s) = \frac{(t^2 - s^2)^{k+1}}{(k+1)!}, \quad t, s \in \mathbb{R}.$$

Actually, we have

$$h_{k+1}(t, s) = \int_s^t h_k(u, s) d_g u$$

$$= 2 \int_s^t \frac{(u^2 - s^2)^k}{k!} u \, du$$

$$= \int_s^t \frac{(u^2 - s^2)^k}{k!} d(u^2 - s^2)$$

$$= \frac{(u^2 - s^2)^{k+1}}{(k+1)!} \Big|_{u=s}^{u=t}$$

$$= \frac{(t^2 - s^2)^{k+1}}{(k+1)!}, \quad s, t \in \mathbb{R}.$$

The above example can be generalized as follows.

Theorem 2.17. *Let $I = [a, b]$ and g be Riemann integrable on $[a, b]$ and g' exists on $[a, b]$. Then,*

$$h_k(t, s) = \frac{(g(t) - g(s))^k}{k!}$$

for $t, s \in I$ and any $k \in \mathbb{N}$.

Proof. We have
$$h_0(t,s) = 1, \quad t, s \in \mathbb{R},$$
$$h_1(t,s) = \int_s^t h_0(u,s) d_g u$$
$$= \int_s^t d_g u$$
$$= g(t) - g(s), \quad t, s \in \mathbb{R},$$
and
$$h_2(t,s) = \int_s^t h_1(u,s) d_g u$$
$$= \int_s^t (g(u) - g(s)) g'(u) du$$
$$= \int_s^t (g(u) - g(s)) g'(u) du$$
$$= \int_s^t (g(u) - g(s)) d(g(u) - g(s))$$
$$= \frac{(g(u) - g(s))^2}{2} \bigg|_{u=s}^{u=t}$$
$$= \frac{(g(t) - g(s))^2}{2}, \quad t, s \in \mathbb{R}.$$

Assume that
$$h_k(t,s) = \frac{(g(t) - g(s))^k}{k!}, \quad t, s \in \mathbb{R},$$
for some $k \in \mathbb{N}$. We will prove that
$$h_{k+1}(t,s) = \frac{(g(t) - g(s))^{k+1}}{(k+1)!}, \quad t, s \in \mathbb{R}.$$
Actually, we have
$$h_{k+1}(t,s) = \int_s^t h_k(u,s) d_g u$$
$$= \int_s^t \frac{(g(u) - g(s))^k}{k!} g'(u) du$$

$$= \int_s^t \frac{(g(u)-g(s))^k}{k!} d(g(u)-g(s))$$

$$= \frac{(g(u)-g(s))^{k+1}}{(k+1)!} \Big|_{u=s}^{u=t}$$

$$= \frac{(g(t)-g(s))^{k+1}}{(k+1)!}, \quad t,s \in \mathbb{R}.$$

By induction arguments, we conclude that the assertion is true for any $t, s \in [a,b]$ and $k \in \mathbb{N}$. This completes the proof. □

Theorem 2.18. Let $t_0 \in I$. For $k, m \in \mathbb{N}_0$, we have

$$h_{k+m+1}(t, t_0) = \int_{t_0}^t h_k(t,s) h_m(s, t_0) d_g s, \quad t \in I.$$

Proof. Let

$$f(t) = \int_{t_0}^t h_k(t,s) h_m(s, t_0) d_g s, \quad t \in I.$$

Then,

$$f_g'(t) = h_k(t,t) h_m(t, t_0)$$

$$+ \int_{t_0}^t h_{kg}'(t,s) h_m(s, t_0) d_g s$$

$$= \int_{t_0}^t h_{k-1}(s) h_m(s, t_0) d_g s$$

$$\vdots$$

$$f_g^{(k)}(t) = \int_{t_0}^t h_m(s, t_0) d_g s$$

$$= h_{m+1}(t, t_0), \quad t \in I.$$

Hence,

$$f_g^{(k-1)}(t) = \int_{t_0}^t h_{m+1}(s, t_0) d_g s$$

$$= h_{m+2}(t, t_0), \quad t \in I,$$

i.e.,
$$f_g^{(k-1)}(t) = h_{m+2}(t, t_0), \quad t \in I.$$

Continuing this, we get
$$f_g'(t) = h_{k+m}(t, t_0), \quad t \in I,$$

whereupon
$$f(t) = \int_{t_0}^{t} h_{k+m}(s, t_0) d_g s$$
$$= h_{k+m+1}(t, t_0), \quad t \in I,$$

i.e.,
$$f(t) = h_{k+m+1}(t, t_0), \quad t \in I.$$

This completes the proof. □

Theorem 2.19. *Let $t_0 \in I$. For $k \in \mathbb{N}_0$, we have*
$$h_k(t, t_0) \geq \frac{(t-t_0)^k}{k!}, \quad k \in \mathbb{N}, \quad t \geq t_0, \quad t \in I. \tag{2.8}$$

Proof. Let
$$f(t) = (t - t_0)^{k+1}, \quad t \in I, \quad t \geq t_0.$$

We have
$$f_g'(t) = \sum_{l=0}^{k} (t-t_0)^{k-l}(t-t_0)^l, \quad t \in I, \quad t \geq t_0.$$

Note that (2.8) holds for $k = 0$ and $k = 1$. Assume that (2.8) holds for some $k \in \mathbb{N}$. We shall prove the inequality for $k + 1$. We have
$$h_{k+1}(t, t_0) = \int_{t_0}^{t} h_k(u, t_0) d_g u$$
$$\geq \frac{1}{k!} \int_{t_0}^{t} (u - t_0)^k d_g u$$

$$= \frac{1}{(k+1)!} \int_{t_0}^{t} \sum_{j=0}^{k} (u-t_0)^k d_g u$$

$$= \frac{1}{(k+1)!} \int_{t_0}^{t} \sum_{j=0}^{k} (u-t_0)^{k-j}(u-t_0)^j d_g u$$

$$= \frac{1}{(k+1)!} \int_{t_0}^{t} f_g'(u) d_g u$$

$$= \frac{1}{(k+1)!} f(u) \Big|_{u=t_0}^{u=t}$$

$$= \frac{(t-t_0)^{k+1}}{(k+1)!}, \quad t \geq t_0, \quad t \in I.$$

By the induction arguments, it follows that (2.8) holds for all $k \in \mathbb{N}$. This completes the proof. □

Now, we shall establish the Stieltjes version of the Taylor Formula.

Theorem 2.20 (**The Stieltjes–Taylor Formula**)**.** *Let $n \in \mathbb{N}$. Suppose that $f: I \to \mathbb{R}$ and $f_g^{(k)}(t_0)$ exist for $k \in \{0, 1, \ldots, n\}$ and for some $t_0 \in I$. Then, there exists a constant $c \in I$ so that*

$$f(t) = f(t_0) + f_g'(t_0) h_1(t, t_0)$$
$$+ \cdots + f_g^{(n)}(t_0) h_n(t, t_0) + f_g^{(n+1)}(c) h_{n+1}(t, t_0), \quad t \in I. \quad (2.9)$$

Proof. First, define a polynomial $P_n(x, x_0)$, $x \in I$ as follows:

$$P_n(t, t_0) = a_0 + a_1 h_1(t, t_0) + a_2 h_2(t, t_0) + \cdots + a_n h_n(t, t_0), \quad t \in I, \quad (2.10)$$

where a_k, $k \in \{0, 1, 2, \ldots, n\}$, are unknowns to be determined. Now, differentiating (2.10), in the sense of Stieltjes, we obtain

$$P_{ng}'(t, t_0) = a_1 + a +_2 h_1(t, t_0) + \cdots + a_n h_{n-1}(t, t_0),$$
$$P_{ng}''(t, t_0) = a_2 + a_3 h_1(t, t_0) + \cdots + a_n h_{n-2}(t, t_0),$$
$$\vdots$$
$$P_{ng}^{(n)}(t, t_0) = a_n, \quad t \in I.$$

Then,
$$P_{ng}^{(n)}(t,t_0) = a_n$$
$$= f_g^{(n)}(t_0)$$
and
$$P_{ng}^{(n-1)}(t_0) = a_{n-1}$$
$$= f_g^{(n-1)}(t_0),$$
and so on. Next,
$$P'_{ng}(t_0,t_0) = a_1$$
$$= f'_g(t_0),$$
and
$$P_n(t_0,t_0) = a_0$$
$$= f(t_0).$$
Therefore,
$$P_n(t,t_0) = f(t_0) + f'_g(t_0)h_1(t,t_0) + \cdots + f_g^{(n)}(t_0)h_n(t,t_0), \quad t \in I. \tag{2.11}$$
Now, let
$$R_n(t) = f(t) - P_n(t,t_0), \quad t \in I.$$
Then differentiating this equation k-times in the sense of Stieltjes, we have
$$R_{ng}^{(k)}(t,t_0) = f_g^{(k)}(t) - P_{ng}^{(k)}(t,t_0), \quad t \in I,$$
and
$$R_{ng}^{(k)}(t_0,t_0) = f_g^{(k)}(t_0) - P_{ng}^{(k)}(t_0,t_0)$$
$$= 0.$$
Moreover,
$$R_{ng}^{(n+1)}(t) = f_g^{(n+1)}(t), \quad t \in I.$$
Next, let
$$\phi(t) = h_{n+1}(t,t_0), \quad t \in I.$$

Then differentiating this equation k-times in the sense of Stieltjes, we obtain

$$\phi_g^{(k)}(t) = h_{n-kg}(t, t_0), \quad t \in I, \quad k \in \{0, 1, \ldots, n-1\},$$

$$\phi_g^{(k)}(t_0) = 0, \quad k \in \{0, 1, \ldots, n\},$$

and

$$\phi_g^{(n+1)}(t_0) = 1.$$

Now, applying the Lagrange mean value theorem (Theorem 2.12), we arrive at

$$\frac{R_n(t)}{\phi(t)} = \frac{R_n(t) - R_n(t_0)}{\phi(t) - \phi(t_0)}$$

$$= \frac{R'_{ng}(t_1)}{\phi'_g(t_1)}$$

$$= \frac{R'_{ng}(t_1) - R'_{ng}(t_0)}{\phi'_g(t_1) - \phi'_g(t_0)}$$

$$= \frac{R''_{ng}(t_2)}{\phi''_g(t_2)}$$

$$\vdots$$

$$= \frac{R_{ng}^{(n+1)}(t_{n+1})}{\phi_g^{(n+1)}(t_0)}$$

$$= f_g^{(n+1)}(t_0).$$

Let

$$c = t_{n+1}.$$

Then

$$R_n(t) = f_g^{(n+1)}(c)\phi(t)$$

$$= f_g^{(n+1)}(c)h_{n+1}(t, t_0), \quad t \in I,$$

whereupon we obtain (2.9). This completes the proof. $\square$

Definition 2.4. The polynomial (2.11) is said to be the Taylor polynomial of f at t_0.

2.7 Advanced Practical Problems

Problem 2.1. Let $I = \mathbb{R}$ and
$$g(t) = (1+t)^3, \quad t \in \mathbb{R}.$$
Find $f'_g(t)$, $t \in \mathbb{R}$, $t \neq -1$, where

(1)
$$f(t) = \frac{1}{t^2 + 2t + 2}, \quad t \in \mathbb{R},$$

(2)
$$f(t) = e^{t^2}, \quad t \in \mathbb{R},$$

(3)
$$f(t) = 3(\cos t)^2 = -2(\cos t)^3, \quad t \in \mathbb{R},$$

(4)
$$f(t) = \cos(\sin t) + \sin(\cos t), \quad t \in \mathbb{R},$$

(5)
$$f(t) = \cos(t^3) + (\cos t)^3, \quad t \in \mathbb{R}.$$

Answer 2.3.

(1)
$$-\frac{2}{3(1+t)(t^2 + 2t + 2)^2}, \quad t \in \mathbb{R}, \quad t \neq -1,$$

(2)
$$\frac{2te^t}{3(1+t)^2}, \quad t \in \mathbb{R}, \quad t \neq -1,$$

(3)
$$\frac{2t\cos(t^2) + \sin(2t)}{3(1+t)^2}, \quad t \in \mathbb{R}, \quad t \neq -1,$$

(4)
$$\frac{-3\sin(2t) + 6(\cos t)^2 \sin t}{3(1+t)^2}, \quad t \in \mathbb{R}, \quad t \neq -1,$$

(5)
$$\frac{-\cos t \sin(\sin t) - \sin t \cos(\cos t)}{3(1+t)^2}, \quad t \in \mathbb{R}, \quad t \neq -1,$$

(6)
$$\frac{-3t^2 \sin(t^3) - 3(\cos t)^2 \sin t}{3(1+t)^2}, \quad t \in \mathbb{R}, \quad t \neq -1.$$

Problem 2.2. Let $I = \mathbb{R}$ and
$$g(t) = \begin{cases} t^2 + 1 & \text{for } t \in (0, \infty), \\ -5 & \text{for } (-\infty, 0], \end{cases}$$

and
$$f(t) = \begin{cases} -3 & \text{for } t \in (0, \infty) \\ t^2 + t & \text{for } (-\infty, 0]. \end{cases}$$

Find $f'_g(0)$.

Answer 2.4. $-\frac{1}{2}$.

Problem 2.3. Let $I = [1, \infty)$,
$$g(t) = 1 + t^2$$

and
$$f(t) = t^3 - t + 1, \quad t \in I.$$

Find $f_g^{(3)}(t)$, $t \in \mathbb{R}$.

Answer 2.5.
$$\frac{-3(t^2 + 10)}{8t^5}, \quad t \in I.$$

Problem 2.4. Let $I = \mathbb{R}$,
$$g(t) = 1 + 4t^5,$$
$$f_1(t) = -1 + 2t - t^3, \quad \text{and}$$
$$f_2(t) = 1 - t + t^2, \quad t \in \mathbb{R}.$$

Check Theorems 2.5 and 2.6.

Chapter 3

Stieltjes Elementary Functions

In this chapter, we introduce Stieltjes regressive functions and explore some of their properties. Functions such as Stieltjes exponential functions, Stieltjes trigonometric functions and Stieltjes hyperbolic functions are defined, and some basic relations are deduced.

Suppose that $I \subseteq \mathbb{R}$ and $g \colon I \to \mathbb{R}$ is monotone nondecreasing function that is continuous from the left everywhere. For a function $f \colon I \to \mathbb{R}$, define the function

$$\Delta f = f(t+) - f(t), \quad t \in I.$$

3.1 Stieltjes Regressive Functions

Definition 3.1. A function $f \colon I \to \mathbb{R}$ is said to be Stieltjes regressive with respect to g provided

$$1 + \Delta g(t) f(t) \neq 0, \quad t \in I.$$

The set of all Stieltjes regressive functions on I is denoted by $\mathscr{R}_g$.

Definition 3.2. In $\mathscr{R}_g$, define the Stieltjes circle plus $\oplus_g$ as

$$(f \oplus_g h)(t) = f(t) + h(t) + \Delta g(t) f(t) h(t), \quad t \in I.$$

Example 3.1. Let $I = [0, \infty)$,

$$g(t) = \begin{cases} 1 & \text{for } t \in [0, 2], \\ 3 & \text{for } t \in (2, \infty), \end{cases}$$

$$f(t) = 1 + t,$$

and

$$h(t) = t^2, \quad t \in I.$$

We will find

$$(f \oplus_g h)(2).$$

We have

$$f(2) = 1 + 2$$
$$= 3,$$
$$h(2) = 2^2$$
$$= 4$$

and

$$\Delta g(2) = g(2+) - g(2)$$
$$= 3 - 1$$
$$= 2.$$

Hence,

$$(f \oplus_g h)(2) = f(2) + h(2) + \Delta g(2) f(2) h(2)$$
$$= 3 + 4 + 2 \cdot 3 \cdot 4$$
$$= 7 + 24$$
$$= 31.$$

Example 3.2. Let $I = (0, 10)$ and define
$$f(t) = t,$$
and
$$g(t) = t^2, \quad t \in I.$$
We will find
$$(f \oplus_g g)(t), \quad t \in I.$$
By definition, we have
$$\Delta g(t) = 0, \quad t \in I,$$
and
$$(f \oplus_g g)(t) = f(t) + g(t) + \Delta g(t) f(t) g(t)$$
$$= t + t^2, \quad t \in I.$$

Exercise 3.1. Let $I = [0, \infty)$ and define
$$g(t) = \begin{cases} -1 & \text{for } t \in [0, 1], \\ 5 & \text{for } t \in (1, \infty), \end{cases}$$
$$f(t) = 2 - t,$$
and
$$h(t) = 1 + t, \quad t \in I.$$
Find
$$(f \oplus_g h)(1).$$

Answer 3.1. 15.

Theorem 3.1. *The set $\mathscr{R}_g$ under the operation $\oplus_g$ is an Abelian group.*

Proof. Let $f, h, l \in \mathscr{R}_g$ be arbitrarily chosen. Then
$$1 + \Delta g(t) f(t) \neq 0,$$
$$1 + \Delta g(t) h(t) \neq 0,$$
and
$$1 + \Delta g(t) l(t) \neq 0, \quad t \in I.$$

Now, we have
$$1 + \Delta g(t)(f \oplus_g h)(t)$$
$$= 1 + \Delta g(t)(f(t) + h(t) + \Delta g(t) f(t) h(t))$$
$$= 1 + \Delta g(t) f(t) + \Delta g(t) h(t) + (\Delta g(t))^2 f(t) h(t)$$
$$= (1 + \Delta g(t) f(t)) + \Delta g(t) h(t)(1 + \Delta g(t) f(t))$$
$$= (1 + \Delta g(t) f(t))(1 + \Delta g(t) h(t))$$
$$\neq 0, \quad t \in I.$$

Thus, $f \oplus_g h \in \mathscr{R}_g$. Next,
$$(f \oplus_g h) \oplus_g l(t) = (f(t) + h(t) + \Delta g(t) f(t) h(t)) + l(t)$$
$$+ \Delta g(t)(f(t) + h(t) + \Delta g(t) f(t) h(t)) l(t)$$
$$= f(t) + h(t) + \Delta g(t) f(t) h(t) + l(t) + \Delta g(t) f(t) l(t)$$
$$+ \Delta g(t) h(t) l(t) + (\Delta g(t))^2 f(t) h(t) l(t)$$
$$= f(t) + (h(t) + l(t) + \Delta g(t) h(t) l(t))$$
$$+ \Delta g(t)(h(t) + g(t) + \Delta g(t) h(t) l(t)) f(t)$$
$$= f(t) + (h \oplus_g l)(t) + \Delta g(t)(h \oplus_g l)(t) f(t)$$
$$= f(t) \oplus_g (h \oplus_g l)(t), \quad t \in I.$$

Therefore, in $(\mathscr{R}_g, \oplus_g)$, the associativity law holds. Note that $0 \in \mathscr{R}_g$ and
$$(0 \oplus_g f)(t) = 0 + f(t) + \Delta g(t) \cdot 0 \cdot f(t)$$
$$= f(t), \quad t \in I.$$

Let
$$m(t) = -\frac{f(t)}{1 + \Delta g(t) f(t)}, \quad t \in I.$$
We will show that $m \in \mathscr{R}_g$. Actually, we have
$$1 + \Delta g(t) m(t) = 1 - \frac{\Delta g(t) f(t)}{1 + \Delta g(t) f(t)}$$
$$= \frac{1}{1 + \Delta g(t) f(t)}$$
$$\neq 0, \quad t \in I.$$
Then,
$$(f \oplus m)(t) = f(t) + m(t) + \Delta g(t) f(t) m(t)$$
$$= f(t) - \frac{f(t)}{1 + \Delta g(t) f(t)} - \frac{\Delta g(t)(f(t))^2}{1 + \Delta g(t) f(t)}$$
$$= \frac{f(t) + \Delta g(t)(f(t))^2 - f(t) - \Delta g(t)(f(t))^2}{1 + \Delta g(t) f(t)}$$
$$= 0, \quad t \in I.$$
So, every element of $\mathscr{R}_g$ has inverse with respect to the Stieltjes circle plus $\oplus_g$. Also,
$$(f \oplus_g h)(t) = f(t) + h(t) + \Delta g(t) f(t) h(t)$$
$$= h(t) + f(t) + \Delta g(t) h(t) f(t)$$
$$= (h \oplus_g f)(t), \quad t \in I,$$
i.e., the commutativity law holds. Therefore, $(\mathscr{R}_g, \oplus_g)$ forms an Abelian group. This completes the proof. $\square$

Definition 3.3. The group $(\mathscr{R}_g, \oplus_g)$ is said to be Stieltjes regressive group.

Definition 3.4. For $f \in \mathscr{R}_g$, define the Stieltjes circle minus as follows:
$$(\ominus_g f)(t) = -\frac{f(t)}{1 + \Delta g(t) f(t)}, \quad t \in I.$$

From the proof of Theorem 4.1, it follows that $\ominus_g f \in \mathscr{R}_g$ provided $f \in \mathscr{R}_g$.

Example 3.3. Let I, g, and f be as in Example 3.1. Then,

$$(\ominus_g f)(2) = -\frac{f(t)}{1 + \Delta g(t) f(t)}$$
$$= -\frac{1+2}{1 + 2(1+2)}$$
$$= -\frac{3}{7}.$$

Next,

$$(\ominus_g g)(2) = -\frac{g(2)}{1 + \Delta g(2) g(2)}$$
$$= -\frac{1}{1 + 2 \cdot 1}$$
$$= -\frac{1}{3}.$$

Example 3.4. Let I, g, and f be as in Example 3.2. Then,

$$(\ominus_g f)(t) = -\frac{f(t)}{1 + \Delta g(t) f(t)}$$
$$= -f(t)$$
$$= -t, \quad t \in I,$$

and

$$(\ominus_g g)(t) = -\frac{g(t)}{1 + \Delta g(t) g(t)}$$
$$= -t^2, \quad t \in I.$$

Exercise 3.2. Let $I = \mathbb{R}$ and define

$$g(t) = \frac{1}{2} t$$

and
$$f(t) = t - t^2, \quad t \in I.$$
Then, find
$$(\ominus_g f)(t), \quad t \in I.$$

Answer 3.2. $-t + t^2$, $t \in I$.

Definition 3.5. For $f, h \in \mathcal{R}_g$, define
$$f \ominus_g h = f \oplus_g (\ominus_g h).$$

By definition, we get
$$(f \ominus_g h)(t) = f(t) + (\ominus_g h)(t) + \Delta g(t) f(t) (\ominus_g h)(t)$$
$$= f(t) - \frac{h(t)}{1 + \Delta g(t) h(t)} - \frac{\Delta g(t) f(t) h(t)}{1 + \Delta g(t) h(t)}$$
$$= \frac{f(t) + \Delta g(t) f(t) h(t) - h(t) - \Delta g(t) f(t) h(t)}{1 + \Delta g(t) h(t)}$$
$$= \frac{f(t) - h(t)}{1 + \Delta g(t) h(t)}.$$

Thus,
$$(f \ominus_g h)(t) = \frac{f(t) - h(t)}{1 + \Delta g(t) h(t)}, \quad t \in I.$$

Example 3.5. Let I, g, f, and h be as in Example 3.1. Then,
$$(f \ominus_g h)(2) = \frac{f(2) - h(2)}{1 + \Delta g(2) h(2)}$$
$$= \frac{3 - 4}{1 + 2 \cdot 4}$$
$$= -\frac{1}{9}.$$

Exercise 3.3. Let $I = \mathbb{R}$ and define

$$g(t) = \frac{1}{4}t,$$
$$f(t) = 2 - t^2,$$

and

$$h(t) = 2 + t^2, \quad t \in I.$$

Then, find

(1) $(f \oplus_g h)(t)$, $t \in I$,
(2) $(\ominus_g f)(t)$, $t \in I$,
(3) $(\ominus_g h)(t)$, $t \in I$,
(4) $(f \ominus_g h)(t)$, $t \in I$,
(5) $(h \ominus_g f)(t)$, $t \in I$.

Theorem 3.2. *Let $f, h \in \mathscr{R}_g$. Then we have the following:*

(1) $\ominus_g(\ominus_g f) = f$,
(2) $f \ominus_g f = 0$,
(3) $f \ominus_g h \in \mathscr{R}_g$,
(4) $\ominus_g(f \ominus_g h) = h \ominus_g f$,
(5) $(\ominus_g(f \oplus_g h)) = (\ominus_g f) \oplus_g (\ominus_g h)$,
(6) $f \oplus_g (\ominus_g h) = f + h$.

Proof. (1) We have

$$(\ominus_g(\ominus_g f))(t) = -\frac{(\ominus_g f)(t)}{1 + \Delta g(t)(\ominus_g f)(t)}$$

$$= \frac{\frac{f(t)}{1+\Delta g(t)f(t)}}{1 - \frac{\Delta g(t)f(t)}{1+\Delta g(t)f(t)}}$$

$$= \frac{\frac{f(t)}{1+\Delta g(t)f(t)}}{\frac{1}{1+\Delta g(t)f(t)}}$$

$$= f(t), \quad t \in I.$$

Thus,
$$(\ominus_g(\ominus_g f))(t) = f(t), \quad t \in I.$$

(2) We have
$$(f \ominus_g f)(t) = (f \oplus_g (\ominus_g f))(t)$$
$$= f(t) + \left(-\frac{f(t)}{1 + \Delta g(t)f(t)}\right) - \Delta g(t)\frac{(f(t))^2}{1 + \Delta g(t)f(t)}$$
$$= \frac{f(t) + \Delta g(t)(f(t))^2 - f(t) - \Delta g(t)(f(t))^2}{1 + \Delta g(t)f(t)}$$
$$= 0, \quad t \in I.$$

Thus,
$$(f \ominus_g f)(t) = 0, \quad t \in I.$$

(3) Since $f, h \in \mathscr{R}_g$, we have that
$$1 + \Delta g(t)f(t) \neq 0, \quad t \in I$$
and
$$1 + \Delta g(t)h(t) \neq 0, \quad t \in I.$$

Then
$$1 + \Delta g(t)(f \ominus_g h)(t) = 1 + \Delta g(t)\frac{f(t) - h(t)}{1 + \Delta g(t)h(t)}$$
$$= \frac{1 + \Delta g(t)h(t) + \Delta g(t)f(t) - \Delta g(t)h(t)}{1 + \Delta g(t)h(t)}$$
$$= \frac{1 + \Delta g(t)f(t)}{1 + \Delta g(t)h(t)}$$
$$\neq 0, \quad t \in I.$$

Thus, $f \ominus_g h \in \mathscr{R}_g$.

(4) We have
$$(h \ominus_g f)(t) = \frac{h(t) - f(t)}{1 + \Delta g(t) f(t)}, \quad t \in I.$$

Hence, using the computations in the previous point, we get

$$(\ominus_g(f \ominus_g h))(t) = -\frac{(f \ominus_g h)(t)}{1 + \Delta g(t)(f \ominus_g h)(t)}$$

$$= -\frac{\frac{f(t)-h(t)}{1+\Delta g(t)h(t)}}{1 + \frac{\Delta g(t)(f(t)-h(t))}{1+\Delta g(t)h(t)}}$$

$$= -\frac{f(t) - h(t)}{1 + \Delta g(t)h(t) + \Delta g(t)(f(t) - h(t))}$$

$$= \frac{h(t) - f(t)}{1 + \Delta g(t) f(t)}$$

$$= (h \ominus_g f)(t), \quad t \in I.$$

Thus,
$$(\ominus_g(f \ominus_g h))(t) = (h \ominus_g f)(t), \quad t \in I.$$

(5) We have

$$(\ominus_g(f \oplus_g h))(t) = -\frac{(f \oplus_g h)(t)}{1 + \Delta g(t)(f \oplus_g h)(t)}$$

$$= -\frac{f(t) + h(t) + \Delta g(t) f(t) h(t)}{1 + \Delta g(t)(f(t) + h(t) + \Delta g(t) f(t) h(t))}$$

$$= -\frac{f(t) + h(t) + \Delta g(t) f(t) h(t)}{1 + \Delta g(t) f(t) + \Delta g(t) h(t)(1 + \Delta g(t) f(t))}$$

$$= -\frac{f(t) + h(t) + \Delta g(t) f(t) h(t)}{(1 + \Delta g(t) f(t))(1 + \Delta g(t) h(t))}, \quad t \in I.$$

Next,

$$((\ominus_g f) \oplus_g (\ominus_g h))(t)$$
$$= (\ominus_g f)(t) + (\ominus_g h)(t) + \Delta g(t)(\ominus_g f)(t)(\ominus_g h)(t)$$

$$= -\frac{f(t)}{1+\Delta g(t)f(t)} - \frac{h(t)}{1+\Delta g(t)h(t)}$$
$$+ \frac{\Delta g(t)f(t)h(t)}{(1+\Delta g(t)f(t))(1+\Delta g(t)h(t))}$$
$$= -\frac{1}{(1+\Delta g(t)f(t))(1+\Delta g(t)h(t))}\bigg(f(t)(1+\Delta g(t)h(t))$$
$$+ h(t)(1+\Delta g(t)f(t)) - \Delta g(t)f(t)h(t)\bigg)$$
$$= -\frac{f(t)+h(t)+\Delta g(t)f(t)h(t)}{(1+\Delta g(t)f(t))(1+\Delta g(t)h(t))}$$
$$= (\ominus_g(f \oplus_g h))(t), \quad t \in I.$$

Thus,
$$(\ominus_g(f \oplus_g h))(t) = ((\ominus_g f) \oplus_g (\ominus_g h))(t), \quad t \in I.$$

(6) We have
$$(f \oplus_g (\ominus_g h))(t) = f(t) \oplus_g \frac{h(t)}{1+\Delta g(t)f(t)}$$
$$= f(t) + \frac{h(t)}{1+\Delta g(t)f(t)} + \Delta g(t)\frac{f(t)h(t)}{1+\Delta g(t)f(t)}$$
$$= f(t) + \frac{1+\Delta g(t)f(t)}{1+\Delta g(t)f(t)}h(t)$$
$$= f(t) + h(t), \quad t \in I.$$

Thus,
$$(f \oplus_g (\ominus_g h))(t) = (f+h)(t), \quad t \in I.$$

This completes the proof. □

Definition 3.6. For $f \in \mathscr{R}_g$, define the Stieltjes circle square as
$$f^g = (-f)(\ominus_g f).$$

For $f \in \mathcal{R}_g$, we have

$$f^g(t) = (-f(t))(\ominus_g f)(t)$$
$$= -f(t)\frac{-f(t)}{1+\Delta g(t)f(t)}$$
$$= \frac{(f(t))^2}{1+\Delta g(t)f(t)}, \quad t \in I.$$

Hence,

$$\frac{(f(t))^2}{f^g(t)} = 1+\Delta g(t)f(t), \quad t \in I.$$

Example 3.6. Let $I, g, f,$ and h be as in Example 3.1. Then, we have the following:

$$f^g(2) = (-f(t))(\ominus_g f)(t)$$
$$= \frac{(f(2))^2}{1+\Delta g(2)f(2)}$$
$$= \frac{3^2}{1+2\cdot 3}$$
$$= \frac{9}{7}$$

and

$$g^g(2) = (-g(2))(\ominus_g g)(2)$$
$$= \frac{(g(2))^2}{1+\Delta g(2)g(2)}$$
$$= \frac{1^2}{1+2\cdot 1}$$
$$= \frac{1}{3}.$$

Exercise 3.4. Let $I = \mathbb{R}$ and define

$$g(t) = \frac{7}{8}t + 1$$

and
$$f(t) = t^3, \quad t \in I.$$

Then find
$$f^g(t), \quad t \in I.$$

Theorem 3.3. *Suppose that $f \in \mathscr{R}_g$. Then we have the following:*

(1) $(\ominus_g f)^g = f^g$,
(2) $(f + (\ominus_g f))(t) = \Delta g(t) f^g(t), \quad t \in I$,
(3) $f \oplus_g f^g = f + f^2$.

Proof. (1) We have
$$(\ominus_g f)(t) = -\frac{f(t)}{1 + \Delta g(t) f(t)}, \quad t \in I,$$

and
$$f^g(t) = \frac{(f(t))^2}{1 + \Delta g(t) f(t)}, \quad t \in I.$$

Hence,
$$(\ominus_g f)^g(t) = \frac{\left(-\frac{f(t)}{1+\Delta g(t)f(t)}\right)^2}{1 - \frac{\Delta g(t)f(t)}{1+\Delta g(t)f(t)}}$$
$$= \frac{\frac{(f(t))^2}{(1+\Delta g(t)f(t))^2}}{\frac{1}{1+\Delta g(t)f(t)}}$$
$$= \frac{(f(t))^2}{1 + \Delta g(t) f(t)}, \quad t \in I.$$

Thus,
$$(\ominus_g f)^g(t) = f^g(t), \quad t \in I.$$

(2) We have
$$f(t) + (\ominus_g f)(t) = f(t) - \frac{f(t)}{1 + \Delta g(t) f(t)}$$
$$= f(t) \left(1 - \frac{1}{1 + \Delta g(t) f(t)}\right)$$

$$= f(t)\left(\frac{1+\Delta g(t)f(t)-1}{1+\Delta g(t)f(t)}\right)$$
$$= \Delta g(t)\frac{(f(t))^2}{1+\Delta g(t)f(t)}$$
$$= \Delta g(t)f^g(t), \quad t \in I.$$

Thus,
$$(f+(\ominus_g f))(t) = \Delta g(t)f^g(t), \quad t \in I.$$

(3) We have
$$(f \oplus_g f^g)(t) = f(t) + f^g(t) + \Delta g(t)f(t)f^g(t)$$
$$= f(t) + \frac{(f(t))^2}{1+\Delta g(t)f(t)} + \frac{\Delta g(t)(f(t))^3}{1+\Delta g(t)f(t)}$$
$$= f(t) + \frac{(1+\Delta g(t)f(t))(f(t))^2}{1+\Delta g(t)f(t)}$$
$$= f(t) + (f(t))^2, \quad t \in I.$$

Thus,
$$(f \oplus_g f^g)(t) = (f+f^2)(t), \quad t \in I.$$

This completes the proof. $\square$

3.2 The Stieltjes Exponential Function

Suppose that $f \colon I \to \mathbb{C}$ is a continuous and Stieltjes regressive function on I and $t_0, T \in I$, $t_0 < T$. Set

$$\widetilde{f}(t) = \begin{cases} f(t) & \text{for } t \in [t_0, T]\backslash D_g, \\ \frac{\log(1+f(t)\Delta g(t))}{\Delta g(t)} & \text{for } t \in [t_0, T] \cap D_g. \end{cases}$$

Definition 3.7. Define the Stieltjes exponential function as follows:
$$e_{f,g}(t,t_0) = \exp\left[\int_{t_0}^t \widetilde{f}(s)d_g s\right], \quad t \in [t_0, T],$$
where $f \in \mathscr{R}_g$.

Example 3.7. Let $I = [0, \infty)$ and
$$g(t) = \begin{cases} -1+t & \text{for } t \in [0,2], \\ 4 & \text{for } t \in (2,8], \end{cases}$$
$$f(t) = t^2 + t, \quad t \in I.$$

Then
$$\widetilde{f}(t) = \begin{cases} f(t) & \text{for } t \in [0,2), \\ \frac{\log(1+f(2)\Delta g(2))}{\Delta g(2)} & \text{for } t = 2, \\ 4 & \text{for } t \in (2,8] \end{cases}$$
$$= \begin{cases} t^2 + t & \text{for } t \in [0,2), \\ \frac{\log(1+6\cdot 3)}{3} & \text{for } t = 2, \\ t^2 + t & \text{for } t \in (2,8] \end{cases}$$
$$= \begin{cases} t^2 + t & \text{for } t \in [0,2), \\ \frac{\log(19)}{3} & \text{for } t = 2, \\ t^2 + t & \text{for } t \in (2,8]. \end{cases}$$

Therefore,
$$e_{f,g}(t,0) = \exp\left[\int_0^t \widetilde{f}(s)d_g s\right]$$
$$= \begin{cases} \exp\left[\int_0^t \widetilde{f}(s)g'(s)ds\right] & \text{for } t \in [0,2), \\ \exp\left[\int_0^t \widetilde{f}(s)g'(s)ds + \widetilde{f}(2)(g(2+)-g(2))\right] & \text{for } t \in [2,8] \end{cases}$$

$$= \begin{cases} \exp\left[\int_0^t (s^2+s)ds\right] & \text{for } t \in [0,2), \\ \exp\left[\int_0^t (s^2+s)ds + \frac{\log(19)}{3}\cdot 3\right] & \text{for } t \in (2,8] \end{cases}$$

$$= \begin{cases} \exp\left[\left(\frac{s^3}{3}+\frac{s^2}{2}\right)\Big|_{s=0}^{s=t}\right] & \text{for } t \in [0,2), \\ \exp\left[\left(\frac{s^3}{3}+\frac{s^2}{2}\right)\Big|_{s=0}^{s=t} + \log(19)\right] & \text{for } t \in (2,8] \end{cases}$$

$$= \begin{cases} e^{\frac{t^3}{3}+\frac{t^2}{2}} & \text{for } t \in [0,2), \\ 19 e^{\frac{t^3}{3}+\frac{t^2}{2}} & \text{for } t \in [2,8]. \end{cases}$$

Example 3.8. Let $I = [0, \infty)$,

$$f(t) = 2 + t, \quad t \in I,$$

and

$$g(t) = \begin{cases} -2 & \text{for } t = 0, \\ -1 & \text{for } t \in (0,1), \\ 0 & \text{for } t = 1, \\ 1 & \text{for } t \in (1,2), \\ 2 & \text{for } t = 2, \\ 3 & \text{for } t \in (2,3), \\ 5 & \text{for } t \in [3, \infty). \end{cases}$$

We will find $e_{f,g}(3,0)$. We have

$$\begin{aligned} \Delta g(0) &= g(0+) - g(0) \\ &= -1 - (-2) \\ &= 1, \end{aligned}$$

$$\Delta g(1) = g(1+) - g(1)$$
$$= 1 - 0$$
$$= 1,$$
$$\Delta g(2) = g(2+) - g(2)$$
$$= 3 - 2$$
$$= 1,$$
$$\Delta g(3) = g(3+) - g(3)$$
$$= 5 - 5$$
$$= 0,$$

and

$$f(0) = 2 + 0$$
$$= 2,$$
$$f(1) = 2 + 1$$
$$= 3,$$
$$f(2) = 2 + 2$$
$$= 4,$$
$$f(3) = 2 + 3$$
$$= 5,$$

and

$$\widetilde{f}(0) = \frac{\log(1 + f(0)\Delta g(0))}{\Delta g(0)}$$
$$= \frac{\log(1 + 2 \cdot 1)}{1}$$
$$= \log 3,$$
$$\widetilde{f}(1) = \frac{\log(1 + f(1)\Delta g(1))}{\Delta g(1)}$$
$$= \frac{\log(1 + 3 \cdot 1)}{1}$$
$$= \log 4,$$
$$= 2\log 2,$$

$$\widetilde{f}(2) = \frac{\log(1 + f(2)\Delta g(2))}{\Delta g(2)}$$
$$= \frac{\log(1 + 4 \cdot 1)}{1}$$
$$= \log 5,$$
$$\widetilde{f}(3) = f(3)$$
$$= 2 + 3$$
$$= 5.$$

Therefore,

$$e_{f,g}(3, 0) = \exp\left[\int_0^3 \widetilde{f}(s)d_g s\right]$$
$$= \exp\left[\widetilde{f}(0)\Delta g(0) + \widetilde{f}(1)\Delta g(1) + \widetilde{f}(2)\Delta g(2) + \widetilde{f}(3)\Delta g(3)\right]$$
$$= \exp\left[\log 3 + 2\log 2 + \log 5 + 0\right]$$
$$= \exp\left[\log 3 + \log 4 + \log 5\right]$$
$$= \exp\left[\log 60\right]$$
$$= 60.$$

Exercise 3.5. Let $I = [0, \infty)$ and
$$f(t) = 1 + 3t, \quad t \in I,$$
and
$$g(t) = \begin{cases} 0 & \text{for } t = 0, \\ 2 & \text{for } t \in (0, 1), \\ 4 & \text{for } t = 1, \\ 5 & \text{for } t \in (1, 2), \\ 8 & \text{for } t = 2, \\ 9 & \text{for } t \in (2, 4), \\ 15 & \text{for } t \in [4, \infty). \end{cases}$$

Find $e_{f,g}(4, 0)$.

Answer 3.3. 120.

In the following, we deduct some of the properties of the Stieltjes exponential function.

Theorem 3.4. *Suppose $f, h\colon I \to \mathbb{C}$ are continuous functions and Stieltjes regressive and $t_0 \in I$. Then we have the following:*

(1) $e_{f,g}(t_0, t_0) = 1$,

(2) $(e_{f,g})'_g(t, t_0) = f(t) e_{f,g}(t, t_0)$, $t \in I$,

(3) $\frac{1}{e_{f,g}(t,t_0)} = e_{\ominus_g f, g}(t, t_0)$, $t \in I$,

(4) $e_{f,g}(t, t_0) e_{h,g}(t, t_0) = e_{f \oplus_g h}(t, t_0)$, $t \in I$,

(5) $\frac{e_{f,g}(t,t_0)}{e_{h,g}(t,t_0)} = e_{f \ominus_g h, g}(t, t_0)$, $t \in I$.

Proof. (1) We have

$$e_{f,g}(t_0, t_0) = \exp\left[\int_{t_0}^{t_0} \tilde{f}(s) d_g s\right]$$
$$= 1.$$

Thus,

$$e_{f,g}(t_0, t_0) = 1.$$

(2) We have

$$(e_{f,g})'_g(t, t_0) = \left(\exp\left[\int_{t_0}^{t} \tilde{f}(s) d_g s\right]\right)'_g$$

$$= e_{f,g}(t, t_0) \left(\int_{[t_0,t)\setminus D_g} f(s) d_g s + \int_{[t_0,t)\cap D_g} \frac{\log(1 + f(s)\Delta g(s))}{\Delta g(s)} d_g s\right)'_g$$

$$= e_{f,g}(t, t_0) \left(\int_{[t_0,t)\setminus D_g} f(s) d_g s + \sum_{t_k < t} \frac{\log(1 + f(t_k)\Delta g(t_k))}{\Delta g(t_k)} \Delta g(t_k)\right)'_g$$

$$= e_{f,g}(t, t_0) \left(\int_{[t_0,t)\setminus D_g} f(s) d_g s + \sum_{t_k < t} \log(1 + f(t_k)\Delta g(t_k))\right)'_g$$

$$= f(t) e_{f,g}(t, t_0), \quad t \in I$$

where t_k denotes the points of discontinuity of g on I. Thus,
$$(e_{f,g})'_g(t,t_0) = f(t)e_{f,g}(t,t_0), \quad t \in I,$$

(3) We have

$$\frac{1}{e_{f,g}(t,t_0)}$$

$$= \frac{1}{\exp\left[\int_{t_0}^t \widetilde{f}(s)d_g s\right]}$$

$$= \exp\left[-\int_{t_0}^t \widetilde{f}(s)d_g s\right]$$

$$= \exp\left[-\int_{[t_0,t)\setminus D_g} f(s)d_g s - \int_{[t_0,t)\cap D_g} \widetilde{f}(s)d_g s\right]$$

$$= \exp\left[\int_{[t_0,t)\setminus D_g} \ominus_g f(s)d_g s - \sum_{t_k<t} \frac{\log(1+f(t_k)\Delta g(t_k))}{\Delta g(t_k)}\Delta g(t_k)\right]$$

$$= \exp\left[\int_{[t_0,t)\setminus D_g} \ominus_g f(s)d_g s - \sum_{t_k<t} \frac{\log\left(\frac{1}{1+f(t_k)\Delta g(t_k)}\right)}{\Delta g(t_k)}\Delta g(t_k)\right]$$

$$= \exp\left[\int_{[t_0,t)\setminus D_g} \ominus_g f(s)d_g s - \sum_{t_k<t} \frac{\log\left(1+(\ominus_g f(t_k))\Delta g(t_k)\right)}{\Delta g(t_k)}\Delta g(t_k)\right]$$

$$= \exp\left[\int_{[t_0,t)\setminus D_g} \ominus_g f(s)d_g s + \int_{[t_0,t)\cap D_g} \log(1+(\ominus_g f)(s)\Delta g(s))d_g s\right]$$

$$= \exp\left[\int_{t_0}^t \ominus_g \widetilde{f}(s)d_g s\right]$$

$$= e_{\ominus_g f,g}(t,t_0), \quad t \in I$$

where t_k denotes the points of discontinuity of g on I. Thus,

$$\frac{1}{e_{f,g}(t,t_0)} = e_{\ominus_g f,g}(t,t_0), \quad t \in I.$$

(4) We have

$$e_{f,g}(t,t_0)e_{h,g}(t,t_0)$$
$$= \left(\exp\left[\int_{[t_0,t)\backslash D_g} f(s)d_g s + \int_{[t_0,t)\cap D_g} \frac{\log(1+f(s)\Delta g(s))}{\Delta g(s)}d_g s\right]\right)$$
$$\times \left(\exp\left[\int_{[t_0,t)\backslash D_g} h(s)d_g s + \int_{[t_0,t)\cap D_g} \frac{\log(1+h(s)\Delta g(s))}{\Delta g(s)}d_g s\right]\right)$$
$$= \exp\left[\int_{[t_0,t)\backslash D_g} (f(s)+h(s))d_g s\right.$$
$$\left. + \int_{[t_0,t)\cap D_g} \frac{\log(1+f(s)\Delta g(s))+\log(1+h(s)\Delta g(s))}{\Delta g(s)}d_g s\right]$$
$$= \exp\left[\int_{[t_0,t)\backslash D_g} (f\oplus_g h)(s)d_g s\right.$$
$$\left. + \int_{[t_0,t)\cap D_g} \frac{\log(1+(f\oplus_g h)(s)\Delta g(s))}{\Delta g(s)}d_g s\right]$$
$$= e_{(f\oplus_g h),g}(t,t_0), \quad t\in I.$$

Thus,
$$e_{f,g}(t,t_0)e_{h,g}(t,t_0) = e_{(f\oplus_g h),g}(t,t_0), \quad t\in I.$$

(5) Using the identities in points (3) and (4), we find
$$\frac{e_{f,g}(t,t_0)}{e_{h,g}(t,t_0)} = e_{f,g}(t,t_0)e_{\ominus h,g}(t,t_0)$$
$$= e_{f\oplus_g(\ominus_g h),g}(t,t_0)$$
$$= e_{f\ominus_g h,g}(t,t_0), \quad t\in I.$$

Thus,
$$\frac{e_{f,g}(t,t_0)}{e_{h,g}(t,t_0)} = e_{f\ominus_g h,g}(t,t_0), \quad t\in I.$$

This completes the proof. □

3.3 Stieltjes Trigonometric Functions

In this section, we define the Stieltjes trigonometric functions and deduct some of their properties. Let $I \subseteq \mathbb{R}$ and $f : I \to \mathbb{R}$ be the Stieltjes regressive function.

Definition 3.8. Define the Stieltjes trigonometric functions as follows:

$$\sin_{f,g}(t, t_0) = \frac{e_{if,g}(t, t_0) - e_{-if,g}(t, t_0)}{2i}$$

and

$$\cos_{f,g}(t, t_0) = \frac{e_{if,g}(t, t_0) + e_{-if,g}(t, t_0)}{2}, \quad t \in I.$$

Some basic properties of the Stieltjes trigonometric functions are deduced in the following theorem.

Theorem 3.5. *The above Stieltjes trigonometric functions have the following properties*:

(1) $(\sin_{f,g})'_g(t, t_0) = f(t)\cos_{f,g}(t, t_0), \quad t \in I,$
(2) $(\cos_{f,g})'_g(t, t_0) = -f(t)\sin_{f,g}(t, t_0), \quad t \in I,$
(3) $(\cos_{f,g}(t, t_0))^2 + (\sin_{f,g}(t, t_0))^2 = e_{if,g}(t, t_0)e_{-if,g}(t, t_0), \quad t \in I,$
(4) $\cos_{f,g}(t, t_0) + i\sin_{f,g}(t, t_0) = e_{if,g}(t, t_0), \quad t \in I.$

Proof. (1) We have

$$(\sin_{f,g})'_g(t, t_0) = \frac{(e_{if,g})'_g(t, t_0) - (e_{-if,g})'_g(t, t_0)}{2i}$$

$$= \frac{if(t)e_{if,g}(t, t_0) + if(t)e_{-if,g}(t, t_0)}{2i}$$

$$= f(t)\frac{e_{if,g}(t, t_0) + e_{-if,g}(t, t_0)}{2}$$

$$= f(t)\cos_{f,g}(t, t_0), \quad t \in I.$$

Thus,

$$(\sin_{f,g})'_g(t, t_0) = f(t)\cos_{f,g}(t, t_0), \quad t \in I.$$

(2) We have

$$(\cos_{f,g})'_g(t,t_0) = \frac{(e_{if,g})'_g(t,t_0) + (e_{-if,g})'_g(t,t_0)}{2}$$
$$= \frac{if(t)e_{if,g}(t,t_0) - if(t)e_{-if,g}(t,t_0)}{2}$$
$$= -f(t)\frac{e_{if,g}(t,t_0) - e_{-if,g}(t,t_0)}{2i}$$
$$= -f(t)\sin_{f,g}(t,t_0), \quad t \in I.$$

Thus,

$$(\cos_{f,g})'_g(t,t_0) = -f(t)\sin_{f,g}(t,t_0).$$

(3) We have

$$(\cos_{f,g}(t,t_0))^2 + (\sin_{f,g}(t,t_0))^2$$
$$= \left(\frac{e_{if,g}(t,t_0) + e_{-if,g}(t,t_0)}{2}\right)^2 + \left(\frac{e_{if,g}(t,t_0) - e_{-if,g}(t,t_0)}{2i}\right)^2$$
$$= \frac{(e_{if,g}(t,t_0))^2 + 2e_{if,g}(t,t_0)e_{-if,g}(t,t_0) + (e_{-if,g}(t,t_0))^2}{4}$$
$$- \frac{(e_{if,g}(t,t_0))^2 - 2e_{if,g}(t,t_0)e_{-if,g}(t,t_0) + (e_{-if,g}(t,t_0))^2}{4}$$
$$= e_{if,g}(t,t_0)e_{-if,g}(t,t_0), \quad t \in I.$$

Thus,

$$(\cos_{f,g}(t,t_0))^2 + (\sin_{f,g}(t,t_0))^2 = e_{if,g}(t,t_0)e_{-if,g}(t,t_0), \quad t \in I.$$

(4) We have

$$\cos_{f,g}(t,t_0) + i\sin_{f,g}(t,t_0)$$
$$= \frac{e_{if,g}(t,t_0) + e_{-if,g}(t,t_0)}{2} + i\frac{e_{if,g}(t,t_0) - e_{-if,g}(t,t_0)}{2i}$$
$$= \frac{e_{if,g}(t,t_0) + e_{-if,g}(t,t_0) + e_{if,g}(t,t_0) - e_{-if,g}(t,t_0)}{2}$$
$$= e_{if,g}(t,t_0), \quad t \in I.$$

Thus,
$$\cos_{f,g}(t,t_0) + i\sin_{f,g}(t,t_0) = e_{if,g}(t,t_0), \quad t \in I. \qquad \square$$

Exercise 3.6. Prove that
$$\cos_{f,g}(t,t_0) - i\sin_{f,g}(t,t_0) = e_{-if,g}(t,t_0), \quad t \in I.$$

3.4 Stieltjes Hyperbolic Functions

In this section, we will define the Stieltjes hyperbolic functions and deduct some of their properties. Let $I \subseteq \mathbb{R}$ and $f: I \to \mathbb{R}$ be the Stieltjes regressive function.

Definition 3.9. Define the Stieltjes hyperbolic functions as follows:
$$\sinh_{f,g}(t,t_0) = \frac{e_{f,g}(t,t_0) - e_{-f,g}(t,t_0)}{2}, \quad t \in I$$
and
$$\cosh_{f,g}(t,t_0) = \frac{e_{f,g}(t,t_0) + e_{-f,g}(t,t_0)}{2}, \quad t \in I.$$

Now, some basic properties of the Stieltjes hyperbolic functions are deduced in the following theorem.

Theorem 3.6. *The above Stieltjes hyperbolic functions have the following properties:*

(1) $(\sinh_{f,g})'_g(t,t_0) = f(t)\cosh_{f,g}(t,t_0),\ t \in I$,
(2) $(\cosh_{f,g})'_g(t,t_0) = f(t)\sinh_{f,g}(t,t_0),\ t \in I$,
(3) $(\cosh_{f,g}(t,t_0))^2 - (\sinh_{f,g}(t,gt_0))^2 = e_{f,g}(t,t_0)e_{-f,g}(t,t_0),\ t \in I$,
(4) $\cosh_{f,g}(t,t_0) + \sinh_{f,g}(t,t_0) = e_{f,g}(t,t_0),\ t \in I$.

Proof. (1) We have
$$(\sinh_{f,g})'_g(t,t_0) = \frac{(e_{f,g})'_g(t,t_0) - (e_{-f,g})'_g(t,t_0)}{2}$$
$$= \frac{f(t)e_{f,g}(t,t_0) + f(t)e_{-f,g}(t,t_0)}{2}$$

$$= f(t) \frac{e_{f,g}(t, t_0) + e_{-f,g}(t, t_0)}{2}$$
$$= f(t) \cosh_{f,g}(t, t_0), \quad t \in I.$$

Thus,
$$(\sinh_{f,g})'_g(t, t_0) = f(t) \cosh_{f,g}(t, t_0), \quad t \in I.$$

(2) We have
$$(\cosh_{f,g})'_g(t, t_0) = \frac{(e_{f,g})'_g(t, t_0) + (e_{-f,g})'_g(t, t_0)}{2}$$
$$= \frac{f(t) e_{f,g}(t, t_0) - f(t) e_{-f,g}(t, t_0)}{2}$$
$$= f(t) \frac{e_{f,g}(t, t_0) - e_{-f,g}(t, t_0)}{2}$$
$$= f(t) \sinh_{f,g}(t, t_0), \quad t \in I.$$

Thus,
$$(\cosh_{f,g})'_g(t, t_0) = f(t) \sinh_{f,g}(t, t_0), \quad t \in I.$$

(3) We have
$$(\cosh_{f,g}(t, t_0))^2 - (\sinh_{f,g}(t, t_0))^2$$
$$= \left(\frac{e_{f,g}(t, t_0) + e_{-f,g}(t, t_0)}{2}\right)^2 - \left(\frac{e_{f,g}(t, t_0) - e_{-f,g}(t, t_0)}{2}\right)^2$$
$$= \frac{(e_{f,g}(t, t_0))^2 + 2 e_{f,g}(t, t_0) e_{-f,g}(t, t_0) + (e_{-f,g}(t, t_0))^2}{4}$$
$$- \frac{(e_{f,g}(t, t_0))^2 - 2 e_{f,g}(t, t_0) e_{-f,g}(t, t_0) + (e_{-f,g}(t, t_0))^2}{4}$$
$$= e_{f,g}(t, t_0) e_{-f,g}(t, t_0), \quad t \in I.$$

Thus,
$$(\cosh_{f,g}(t, t_0))^2 - (\sinh_{f,g}(t, t_0))^2 = e_{f,g}(t, t_0) e_{-f,g}(t, t_0), \quad t \in I.$$

(4) We have

$$\cosh_{f,g}(t,t_0) + \sinh_{f,g}(t,t_0)$$
$$= \frac{e_{f,g}(t,t_0) + e_{-f,g}(t,t_0)}{2} + \frac{e_{f,g}(t,t_0) - e_{-f,g}(t,t_0)}{2}$$
$$= \frac{e_{f,g}(t,t_0) + e_{-f,g}(t,t_0) + e_{f,g}(t,t_0) - e_{-f,g}(t,t_0)}{2}$$
$$= e_{f,g}(t,t_0), \quad t \in I.$$

Thus,

$$\cosh_{f,g}(t,t_0) + \sinh_{f,g}(t,t_0) = e_{f,g}(t,t_0), \quad t \in I. \qquad \square$$

Exercise 3.7. Prove that

$$\cosh_{f,g}(t,t_0) - \sinh_{f,g}(t,t_0) = e_{-f,g}(t,t_0), \quad t \in I.$$

3.5 Advanced Practical Problems

Problem 3.1. Let $I = [0, \infty)$ and define

$$f(t) = 3 + 2t, \quad t \in I,$$

and

$$g(t) = \begin{cases} -1 & \text{for } t = 0, \\ 3 & \text{for } t \in (0,1), \\ 5 & \text{for } t = 1, \\ 9 & \text{for } t \in (1,2), \\ 15 & \text{for } t = 2, \\ 16 & \text{for } t \in (2,3), \\ 17 & \text{for } t \in [3,\infty). \end{cases}$$

Then, find $e_{f,g}(3,0)$.

Answer 3.4. 2184.

Problem 3.2. Let $I = [0, \infty)$ and define
$$g(t) = 3t + 1,$$
$$f(t) = 1 - 2t,$$
$$h(t) = 2 + 3t, \quad t \in I.$$

Then find

(1)
$$(f \oplus_g h)(t), \quad t \in I,$$

(2)
$$(\ominus_g f)(t), \quad t \in I,$$

(3)
$$(\ominus_g h)(t), \quad t \in I,$$

(4)
$$(f \ominus_g h)(t), \quad t \in I,$$

(5)
$$(h \ominus_g f)(t), \quad t \in I.$$

Problem 3.3. Let $I = [0, 4]$ and define
$$g(t) = t + 2, \quad t \in I$$
and
$$f(t) = 2t, \quad t \in I.$$

Then find

(1) $e_{f,g}(t, 0)$, $t \in I$,
(2) $e_{f \oplus_g g, g}(t, 0)$, $t \in I$.

Problem 3.4. Let $I = [0, \infty)$ and define
$$f(t) = t + 7, \quad t \in I,$$
and
$$g(t) = \begin{cases} -3 & \text{for } t = 0, \\ -2 & \text{for } t \in (0, 1), \\ -1 & \text{for } t = 1, \\ 0 & \text{for } t \in (1, 2), \\ 1 & \text{for } t = 2, \\ 3 & \text{for } t \in (2, 3), \\ 7 & \text{for } t \in [3, \infty). \end{cases}$$

Then find

(1) $\sinh_{f,g}(t, 0)$, $t \in I$,

(2) $\cosh_{f,g}(t, 0)$, $t \in I$.

Problem 3.5. Let $I = [0, \infty)$ and define
$$f(t) = t^2 - 1, \quad t \in I,$$
and
$$g(t) = \begin{cases} 2 & \text{for } t = 0, \\ 4 & \text{for } t \in (0, 1), \\ 6 & \text{for } t = 1, \\ 8 & \text{for } t \in (1, 2), \\ 10 & \text{for } t = 2, \\ 12 & \text{for } t \in (2, 3), \\ 14 & \text{for } t \in [3, \infty). \end{cases}$$

Then find
$$\cosh_{f,g}(t, 0) - 3\sinh_{f,g}(t, 0), \quad t \in I.$$

Chapter 4

The Stieltjes–Laplace Transform

In this chapter, we define the Stieltjes–Laplace transform and prove some of its properties. The Stieltjes–Laplace transforms for some elementary functions are deduced. Formulas for the Stieltjes–Laplace transform of the Stieltjes derivative of an arbitrary order and the Stieltjes–Laplace transform of the Stieltjes integral are presented.

Let $I \subseteq \mathbb{R}$ be such that $\sup I = \infty$. The set of all Stieltjes integrable functions over each compact subinterval of $[t_0, \infty)$, $t_0 \in I$, will be denoted by $V([t_0, \infty))$.

4.1 Functions of Exponential Orders

Definition 4.1. A function f defined on $[0, \infty)$ is said to be of exponential order $\lambda \in \mathbb{R}$ provided $\lambda \in \mathscr{R}_g$, and there is a constant $M > 0$ such that
$$|f(t)| \leq M e_{\lambda,g}(t, t_0) \quad \text{for } t \in [t_0, \infty).$$

Theorem 4.1. *For $z, x \in \mathscr{R}_g$ with $z = x + iy$, $x, y \in \mathbb{R}$, we have*
$$|e_{\ominus_g z, g}(t, t_0)| \leq e_{\ominus_g x, g}(t, t_0) \quad \text{for } t \in [t_0, \infty).$$

Proof. Note that

$$x \oplus_g \frac{iy}{1+x\Delta g(t)} = x + \frac{iy}{1+x\Delta g(t)} + \Delta g(t)\frac{ixy}{1+x\Delta g(t)}$$

$$= x + \frac{iy(1+x\Delta g(t))}{1+x\Delta g(t)}$$

$$= x + iy, \quad t \in [t_0, \infty).$$

Then

$$e_{z,g}(t,t_0) = e_{x+iy,g}(t,t_0)$$

$$= e_{x \oplus_g \frac{iy}{1+x\Delta g(t)}}(t,t_0)$$

$$= e_{x,g}(t,t_0) e_{\frac{iy}{1+x\Delta g(t)}}(t,t_0), \quad t \in [t_0, \infty).$$

Let

$$f(t) = \frac{iy}{1+x\Delta g(t)}, \quad t \in I.$$

Then

$$\widetilde{f}(t) = \begin{cases} iy & \text{if } g \text{ is continuous at } t, \\ \log\left(1 + \frac{iy\Delta g(t)}{1+x\Delta g(t)}\right) & \text{if } g \text{ is discontinuous at } t. \end{cases}$$

Hence,

$$|\widetilde{f}(t)| = \begin{cases} |y| & \text{if } g \text{ is continuous at } t, \\ \left|\log\left(1 + \frac{iy\Delta g(t)}{1+x\Delta g(t)}\right)\right| & \text{if } g \text{ is discontinuous at } t, \end{cases}$$

$$= \begin{cases} |y| & \text{if } g \text{ is continuous at } t, \\ \log\left|1 + \frac{iy\Delta g(t)}{1+x\Delta g(t)}\right| & \text{if } g \text{ is discontinuous at } t, \end{cases}$$

$$\geq \begin{cases} |y| & \text{if } g \text{ is continuous at } t, \\ 0 & \text{if } g \text{ is discontinuous at } t, \end{cases}$$

$$\geq 0, \quad t \in I.$$

Therefore

$$\left|e_{\frac{iy}{1+x\Delta g(t)},g}(t,t_0)\right| \geq 1 \quad \text{for } t \in [t_0, \infty).$$

Hence

$$\left|e_{z,g}(t,t_0)\right| = \left|e_{x,g}(t,t_0)e_{\frac{iy}{1+x\Delta g(t)},g}(t,t_0)\right|$$

$$= \left|e_{x,g}(t,t_0)\right|\left|e_{\frac{iy}{1+x\Delta g(t)},g}(t,t_0)\right|$$

$$\geq \left|e_{x,g}(t,t_0)\right| \quad \text{for } t \in [t_0, \infty)$$

and

$$\left|e_{\ominus_g z,g}(t,t_0)\right| \leq \left|e_{\ominus_g x,g}(t,t_0)\right|$$

$$= e_{\ominus_g x,g}(t,t_0) \quad \text{for } t \in [t_0, \infty).$$

This completes the proof. □

4.2 Definition and Properties of the Stieltjes–Laplace Transform

Definition 4.2. Suppose $f \in V([t_0, \infty))$. Then the Stieltjes–Laplace transform of f is defined by

$$\mathscr{L}_g(f)(z) = \int_{t_0}^{\infty} f(t)\frac{1}{1+\Delta g(t)z}e_{\ominus_g z,g}(t,t_0)d_g t, \qquad (4.1)$$

where $z \in \mathscr{R}_g$ for which the integral (4.1) exists.

Theorem 4.2. *Let $f \in V([t_0, \infty))$ be of exponential order $\lambda \in \mathbb{R}$ and $\lambda \in \mathscr{R}_g$. Then the Stieltjes integral (4.1) converges absolutely for $x > \lambda$, $1 + \Delta g(t)x > 0$, provided that*

$$\lim_{t \to \infty} e_{\lambda \ominus_g x,g}(t,t_0) = 0, \quad z = x + iy.$$

Proof. Let $z = x+iy$, where $x, y \in \mathbb{R}$ and $x > \lambda$, and $1+\Delta g(t)x > 0$. Since f is of exponential order λ, there is a constant $M > 0$ such that
$$|f(t)| \leq M e_{\lambda,g}(t, t_0) \quad \text{for } t \in [t_0, \infty).$$
Now, applying Theorem 4.1, we have
$$\left| \int_{t_0}^{\infty} f(t) e_{\ominus_g z, g}(t, t_0) d_g t \right| \leq \int_{t_0}^{\infty} |f(t)| \frac{1}{|1 + \Delta g(t) z|} |e_{\ominus_g z, g}(t, t_0)| d_g t$$
$$\leq \int_{t_0}^{\infty} |f(t)| \frac{1}{1 + \Delta g(t) x} e_{\ominus_g x, g}(t, t_0) d_g t$$
$$\leq M \int_{t_0}^{\infty} e_{\lambda, g}(t, t_0) \frac{1}{1 + \Delta g(t) x} e_{\ominus_g x, g}(t, t_0) d_g t$$
$$= M \int_{t_0}^{\infty} \frac{1}{1 + \Delta g(t) x} e_{\lambda, g}(t, t_0) e_{\ominus_g x, g}(t, t_0) d_g t$$
$$= M \int_{t_0}^{x} \frac{1}{1 + \Delta g(t) x} e_{\lambda \ominus_g x, g}(t, t_0) d_g t$$
$$= \frac{M}{\lambda - x} \int_{t_0}^{\infty} \frac{\lambda - x}{1 + \Delta g(t) x} e_{\lambda \ominus_g x, g}(t, t_0) d_g t$$
$$= \frac{M}{\lambda - x} \int_{t_0}^{\infty} (\lambda \ominus_g x)(t, t_0) e_{\lambda \ominus_g x}(t, t_0) d_g t$$
$$= \frac{M}{\lambda - x} \int_{t_0}^{\infty} (e_{\lambda \ominus_g x, g})'_g(t, t_0) d_g t$$
$$= \frac{M}{\lambda - x} \lim_{b \to \infty} \int_{t_0}^{b} (e_{\lambda \ominus_g x, g})'_g(t, t_0) d_g t$$
$$= \frac{M}{\lambda - x} \lim_{b \to \infty} e_{\lambda \ominus_g x, g}(t, t_0) \Big|_{t=t_0}^{t=b}$$
$$= \frac{M}{\lambda - x} \lim_{b \to \infty} \left(e_{\lambda \ominus_g x, g}(b, t_0) - 1 \right)$$
$$= \frac{M}{x - \lambda}.$$
Hence, the Stieltjes integral (4.1) converges absolutely. This completes the proof. □

The Stieltjes–Laplace Transform

Example 4.1. We will find
$$\mathscr{L}_g(1)(z), \quad z \in \mathbb{C}, \quad z \neq 0.$$

Suppose $z \in \mathbb{C}$ with $z \neq 0$. By definition, we have

$$\begin{aligned}
\mathscr{L}_g(1)(z) &= \int_{t_0}^{\infty} \frac{1}{1 + \Delta g(t)z} e_{\ominus_g z, g}(t, t_0) d_g t \\
&= -\frac{1}{z} \int_{t_0}^{\infty} \frac{-z}{1 + \Delta g(t)z} e_{\ominus_g z, g}(t, t_0) d_g t \\
&= -\frac{1}{z} \int_{t_0}^{\infty} \ominus_g z\, e_{\ominus_g z, g}(t, t_0) d_g t \\
&= -\frac{1}{z} \int_{t_0}^{\infty} \ominus_g z\, e_{\ominus_g z, g}(t, t_0) d_g t \\
&= -\frac{1}{z} \int_{t_0}^{\infty} (e_{\ominus_g z, g})'_g(t, t_0) d_g t \\
&= -\frac{1}{z} \lim_{b \to \infty} \int_{t_0}^{b} (e_{\ominus_g z, g})'_g(t, t_0) d_g t \\
&= -\frac{1}{z} \lim_{b \to \infty} e_{\ominus_g z, g}(t, t_0) \Big|_{t=t_0}^{t=b} \\
&= -\frac{1}{z} \lim_{b \to \infty} \left(e_{\ominus_g z, g}(b, t_0) - 1 \right) \\
&= \frac{1}{z} \quad \text{provided that } \lim_{t \to \infty} e_{\ominus_g z, g}(t, t_0) = 0.
\end{aligned}$$

Hence,
$$\mathscr{L}_g(1)(z) = \frac{1}{z}, \quad z \neq 0,$$
provided that $\lim_{t \to \infty} e_{\ominus_g z, g}(t, t_0) = 0$.

Example 4.2. Let $z, \lambda \in \mathscr{R}_g$ be such that $z \in \mathbb{C}$ and $z \neq \lambda$. We will find
$$\mathscr{L}_g(e_{\lambda, g}(t, t_0))(z),$$
provided that $\lim_{t \to \infty} e_{\lambda \ominus_g z, g}(t, t_0) = 0$.

By definition, we have

$$\mathscr{L}_g(e_{\lambda,g}(t,t_0))(z) = \int_{t_0}^{\infty} \frac{1}{1+\Delta g(t)z} e_{\lambda,g}(t,t_0) e_{\ominus_g z}(t,t_0) d_g t$$

$$= \int_{t_0}^{\infty} \frac{1}{1+\Delta g(t)z} e_{\lambda,g}(t,t_0) e_{\ominus_g z,g}(t,t_0) d_g t$$

$$= \int_{t_0}^{\infty} \frac{1}{1+\Delta g(t)z} e_{\lambda \ominus_g z,g}(t,t_0) d_g t$$

$$= \frac{1}{\lambda - z} \int_{t_0}^{\infty} \frac{\lambda - z}{1+\Delta g(t)z} e_{\lambda \ominus_g z,g}(t,t_0) d_g t$$

$$= \frac{1}{\lambda - z} \int_{t_0}^{\infty} (\lambda \ominus_g z)(t,t_0) e_{\lambda \ominus_g z,g}(t,t_0) d_g t$$

$$= \frac{1}{\lambda - z} \int_{t_0}^{\infty} (e_{\lambda \ominus_g z,g})'_g(t,t_0) d_g t$$

$$= \frac{1}{\lambda - z} \lim_{b \to \infty} \int_{t_0}^{b} (e_{\lambda \ominus_g z,g})'_g(t,t_0) d_g t$$

$$= \frac{1}{\lambda - z} \lim_{b \to \infty} e_{\lambda \ominus_g z,g}(t,t_0) \Big|_{t=t_0}^{t=b}$$

$$= \frac{1}{\lambda - z} \lim_{b \to \infty} \left(e_{\lambda \ominus_g z,g}(b,t_0) - e_{\lambda \ominus_g z,g}(t_0,t_0) \right)$$

$$= \frac{1}{z - \lambda} \quad \text{provided that} \quad \lim_{t \to \infty} e_{\ominus_g z,g}(t,t_0) = 0.$$

Hence,

$$\mathscr{L}_g(e_{\lambda,g}(t,t_0))(z) = \frac{1}{z-\lambda}, \text{ provided that } \lim_{t\to\infty} e_{\ominus_g z,g}(t,t_0) = 0.$$

Example 4.3. Let $\lambda, \mu, z \in \mathscr{R}_g$ be such that $z \neq \lambda + \Delta g(t)$, $t \in I$. We will find

$$\mathscr{L}_g \left(e_{\frac{\lambda}{1+\mu \Delta g(t)},g}(t,t_0) e_{\mu,g}(t,t_0) \right)(z),$$

provided that

$$\lim_{t\to\infty} e_{(\lambda+\mu)\ominus_g z,g}(t,t_0) = 0.$$

We have

$$\frac{\lambda}{1+\mu\Delta g(t)} \oplus_g \mu = \frac{\lambda}{1+\mu\Delta g(t)} + \mu + \Delta g(t)\frac{\lambda\mu}{1+\mu\Delta g(t)}$$

$$= \frac{\lambda(1+\mu\Delta g(t))}{1+\mu\Delta g(t)} + \mu$$

$$= \lambda + \mu, \quad t \in I.$$

Now, taking the Stieltjes–Laplace transform and using the previous example, we obtain

$$\mathscr{L}_g\left(e_{\frac{\lambda}{1+\mu\Delta g(t)},g}(t,t_0)e_{\mu,g}(t,t_0)\right)(z) = \mathscr{L}_g\left(e_{\frac{\lambda}{1+\mu\Delta g(t)}\oplus\mu,g}(t,t_0)\right)(z)$$

$$= \mathscr{L}_g\left(e_{\lambda+\mu,g}(t,t_0)\right)$$

$$= \frac{1}{z-(\lambda+\mu)}.$$

Thus,

$$\mathscr{L}_g\left(e_{\frac{\lambda}{1+\mu\Delta g(t)},g}(t,t_0)e_{\mu,g}(t,t_0)\right)(z) = \frac{1}{z-(\lambda+\mu)}$$

provided that $\lim_{t\to\infty} e_{(\lambda+\mu)\ominus_g z,g}(t,t_0) = 0$.

Theorem 4.3. Let $f, h \in V([t_0, \infty))$ and $c_1, c_2 \in \mathbb{C}$. Then,

$$\mathscr{L}_g(c_1 f(t) + c_2 h(t))(z) = c_1 \mathscr{L}_g(f(t))(z) + c_2 \mathscr{L}_g(h(t))(z).$$

Proof. By definition, we have

$$\mathscr{L}_g(c_1 f(t) + c_2 h(t))(z) = \int_{t_0}^{\infty} (c_1 f(t) + c_2 h(t))\frac{1}{1+\Delta g(t)z}$$

$$\times e_{\ominus_g z,g}(t,t_0) d_g t$$

$$= \int_{t_0}^{\infty} c_1 f(t)\frac{1}{1+\Delta g(t)z} e_{\ominus_g z,g}(t,t_0) d_g t$$

$$+ \int_{t_0}^{\infty} c_2 h(t) e_{\ominus_g z,g}(t,t_0) d_g t$$

$$= c_1 \int_{t_0}^{\infty} f(t) e_{\ominus_g z, g}(t, t_0) d_g t$$
$$+ c_2 \int_{t_0}^{\infty} h(t) e_{\ominus_g z, g}(t, t_0) d_g t$$
$$= c_1 \mathscr{L}_g(f(t))(z) + c_2 \mathscr{L}_g(h(t))(z).$$

This completes the proof. □

Example 4.4. We will find $\mathscr{L}_g(\sin_{\alpha,g}(t, t_0))(z)$. Since
$$\sin_{\alpha,g}(t, t_0) = \frac{e_{i\alpha,g}(t, t_0) - e_{-i\alpha,g}(t, t_0)}{2i}, \quad t \in I.$$

We have
$$\mathscr{L}_g(\sin_{\alpha,g}(t, t_0))(z) = \mathscr{L}_g\left(\frac{e_{i\alpha,g}(t, t_0) - e_{-i\alpha,g}(t, t_0)}{2i}\right)$$
$$= \frac{1}{2i}\left(\mathscr{L}_g(e_{i\alpha,g}(t, t_0))(z) - \mathscr{L}_g(e_{-i\alpha,g}(t, t_0))(z)\right)$$
$$= \frac{1}{2i}\left(\frac{1}{z - i\alpha} - \frac{1}{z + i\alpha}\right)$$
$$= \frac{1}{2i}\frac{z + i\alpha - z + i\alpha}{(z - i\alpha)(z + i\alpha)}$$
$$= \frac{1}{2i}\frac{2i\alpha}{z^2 + \alpha^2}$$
$$= \frac{\alpha}{z^2 + \alpha^2}.$$

Hence,
$$\mathscr{L}_g(\sin_{\alpha,g}(t, t_0))(z) = \frac{\alpha}{z^2 + \alpha^2}.$$

Example 4.5. We will find $\mathscr{L}_g(\cos_{\alpha,g}(t, t_0))(z)$. Since
$$\cos_{\alpha,g}(t, t_0) = \frac{e_{i\alpha,g}(t, t_0) + e_{-i\alpha,g}(t, t_0)}{2}, \quad t \in I,$$

we have
$$\mathscr{L}_g(\cos_{\alpha,g}(t, t_0))(z) = \mathscr{L}_g\left(\frac{e_{i\alpha,g}(t, t_0) + e_{-i\alpha,g}(t, t_0)}{2}\right)$$
$$= \frac{1}{2}\left(\mathscr{L}_g(e_{i\alpha,g}(t, t_0))(z) + \mathscr{L}_g(e_{-i\alpha,g}(t, t_0))(z)\right)$$

$$= \frac{1}{2}\left(\frac{1}{z-i\alpha} + \frac{1}{z+i\alpha}\right)$$

$$= \frac{1}{2}\frac{z+i\alpha+z-i\alpha}{(z-i\alpha)(z+i\alpha)}$$

$$= \frac{1}{2}\frac{2z}{z^2+\alpha^2}$$

$$= \frac{z}{z^2+\alpha^2}.$$

Hence,

$$\mathscr{L}_g(\cos_{\alpha,g}(t,t_0))(z) = \frac{z}{z^2+\alpha^2}.$$

Example 4.6. We will find $\mathscr{L}_g(\sinh_{\alpha,g}(t,t_0))(z)$. Since

$$\sinh_{\alpha,g}(t,t_0) = \frac{e_{\alpha,g}(t,t_0) - e_{-\alpha,g}(t,t_0)}{2}, \quad t \in I,$$

we have

$$\mathscr{L}_g(\sinh_{\alpha,g}(t,t_0))(z) = \mathscr{L}_g\left(\frac{e_{\alpha,g}(t,t_0) - e_{-\alpha,g}(t,t_0)}{2}\right)$$

$$= \frac{1}{2}(\mathscr{L}_g(e_{\alpha,g}(t,t_0))(z) - \mathscr{L}_g(e_{-\alpha,g}(t,t_0))(z))$$

$$= \frac{1}{2}\left(\frac{1}{z-\alpha} - \frac{1}{z+\alpha}\right)$$

$$= \frac{1}{2}\frac{z+\alpha-z+\alpha}{(z-\alpha)(z+\alpha)}$$

$$= \frac{1}{2}\frac{2\alpha}{z^2-\alpha^2}$$

$$= \frac{\alpha}{z^2-\alpha^2}.$$

Hence,

$$\mathscr{L}_g(\sinh_{\alpha,g}(t,t_0))(z) = \frac{\alpha}{z^2-\alpha^2}.$$

Example 4.7. We will find $\mathscr{L}_g(\cosh_{\alpha,g}(t,t_0))(z)$. Since

$$\cosh_{\alpha,g}(t,t_0) = \frac{e_{\alpha,g}(t,t_0) + e_{-\alpha,g}(t,t_0)}{2}, \quad t \in I,$$

we have

$$\begin{aligned}
\mathscr{L}_g(\cosh_{\alpha,g}(t,t_0))(z) &= \mathscr{L}_g\left(\frac{e_{\alpha,g}(t,t_0) + e_{-\alpha,g}(t,t_0)}{2}\right) \\
&= \frac{1}{2}\left(\mathscr{L}_g(e_{\alpha,g}(t,t_0))(z) + \mathscr{L}_g(e_{-\alpha,g}(t,t_0))(z)\right) \\
&= \frac{1}{2}\left(\frac{1}{z-\alpha} + \frac{1}{z+\alpha}\right) \\
&= \frac{1}{2}\frac{z+\alpha+z-\alpha}{(z-\alpha)(z+\alpha)} \\
&= \frac{1}{2}\frac{2z}{z^2-\alpha^2} \\
&= \frac{z}{z^2-\alpha^2}.
\end{aligned}$$

Hence,

$$\mathscr{L}_g(\cosh_{\alpha,g}(t,t_0))(z) = \frac{z}{z^2-\alpha^2}.$$

Exercise 4.1. Find

$$\mathscr{L}_g\left(e_{2,g}(t,t_0) - e_{-3,g}(t,t_0)\right)(z).$$

Answer 4.1.

$$\frac{5}{(z-2)(z+3)}.$$

4.3 The Stieltjes–Laplace Transform of Stieltjes Derivative

In this section, we deduct the Stieltjes–Laplace transform of Stieltjes derivatives of arbitrary order in the case when $g \in \mathscr{C}(I)$. In this case, using the fact that $\Delta g(t) = 0$, $t \in I$, for a function $f\colon I \to \mathbb{R}$, the Stieltjes–Laplace transform is given by

$$\mathscr{L}_g(f(t))(z) = \int_{t_0}^{\infty} f(t) e_{-z,g}(t,t_0) d_g t,$$

where $z \in \mathscr{R}_g$.

The main result in this section reads as follows.

Theorem 4.4. *Let $f \in V([t_0, \infty))$ be a function of exponential order λ. Then, for $n \in \mathbb{N}$, we have*

$$\mathscr{L}_g\left(f_g^{(n)}(t)\right)(z) = z^n \mathscr{L}_g(f(t))(z) - \sum_{j=0}^{n-1} z^{n-1-j} f_g^j(t_0) \qquad (4.2)$$

provided that

$$\lim_{t \to \infty} f_g^{(k)}(t) e_{-z,g}(t,t_0) = 0, \quad k \in \{0, \ldots, n-1\}.$$

Proof. We will use induction arguments. For $n = 1$, we have

$$\mathscr{L}_g(f_g'(t))(z) = \int_{t_0}^{\infty} f_g'(t) e_{-z,g}(t,t_0) d_g t.$$

Then, using Stieltjes integration by parts, we obtain

$$\mathscr{L}_g(f_g'(t))(z) = \lim_{b \to \infty} \left(f(t) e_{-z,g}(t,t_0) \Big|_{t=t_0}^{t=b} \right) - z \int_{t_0}^{\infty} f(t) e_{-z,g}(t,t_0) d_g t$$

$$= -f(t_0) + \int_{t_0}^{\infty} f(t) e_{-z,g}(t,t_0) d_g t$$

$$= -f(t_0) + z \int_{t_0}^{\infty} f(t) e_{-z,g}(t,t_0) d_g t$$

$$= -f(t_0) + z \mathscr{L}_g(f(t))(z).$$

Thus, (4.2) holds for $n = 1$. Assume that (4.2) holds for some $n \in \mathbb{N}$. We shall prove the assertion for $n+1$, i.e., we shall prove that

$$\mathscr{L}_g\left(f_g^{(n+1)}(t)\right)(z) = z^{n+1} \mathscr{L}_g(f(t))(z) - \sum_{j=0}^{n} z^{n-j} f_g^{(j)}(t_0)$$

provided that
$$\lim_{t\to\infty} f_g^{(k)}(t)e_{-g,z}(t,t_0) = 0, \quad k \in \{0,\ldots,n\}.$$

Actually, using Stieltjes integration by parts, we have

$$\mathscr{L}_g(f_g^{(n+1)}(t))(z) = \int_{t_0}^{\infty} f_g^{(n+1)}(t)e_{-z,g}(t,t_0)d_g t$$

$$= \lim_{b\to\infty}\left(f_g^{(n)}(t)e_{-z,g}(t,t_0)\Big|_{t=t_0}^{t=b}\right)$$

$$- \int_{t_0}^{\infty}(-z)(t)f_g^{(n)}(t)e_{-z,g}(t,t_0)d_g t$$

$$= -f_g^{(n)}(t_0) + \int_{t_0}^{\infty} f_g^{(n)}(t)e_{-z,g}(t,t_0)d_g t$$

$$= -f_g^{(n)}(t_0) + z\int_{t_0}^{\infty} f_g^{(n)}(t)e_{-z,g}(t,t_0)d_g t$$

$$= -f_g^{(n)}(t_0) + z\mathscr{L}_g\left(f_g^{(n)}(t)\right)(z)$$

$$= -f_g^{(n)}(t_0) + z\left(z^n \mathscr{L}_g(f(t))(z) - \sum_{j=0}^{n-1} z^{n-1-j} f_g^{(j)}(t_0)\right)$$

$$= z^{n+1}\mathscr{L}_g(f(t))(z) - \sum_{j=0}^{n} z^{n-j} f_g^{(j)}(t_0).$$

Thus, (4.2) holds for $n+1 \in \mathbb{N}$. Now, applying the principle of mathematical induction, we conclude that (4.2) holds for all $n \in \mathbb{N}$. This completes the proof. □

Example 4.8. Let
$$f(t) = \sin_{\alpha,g}(t,t_0), \quad t \in I.$$

Then, we have
$$f'_g(t) = \alpha \cos_{\alpha,g}(t,t_0), \quad t \in I.$$

Now, by the properties of the Stieltjes–Laplace transform, we get
$$\begin{aligned}\mathscr{L}_g(f'_g(t)) &= \mathscr{L}_g(\alpha \cos_{\alpha,g}(t,t_0))(z) \\ &= \alpha \mathscr{L}_g(\cos_{\alpha,g}(t,t_0))(z) \\ &= \frac{\alpha z}{z^2 + \alpha^2}.\end{aligned}$$
On the other hand, using Theorem 4.4, we find
$$\begin{aligned}\mathscr{L}_g(f'_g(t))(z) &= z\mathscr{L}_g(f(t))(z) - f(t_0) \\ &= z\mathscr{L}_g(\sin_{\alpha,g}(t,t_0))(z) \\ &= \frac{\alpha z}{z^2 + \alpha^2}.\end{aligned}$$

Example 4.9. Let
$$f(t) = \cosh_{\alpha,g}(t,t_0), \quad t \in I.$$
Then, we have
$$\begin{aligned}f'_g(t) &= \alpha \sinh_{\alpha,g}(t,t_0), \\ f''_g(t) &= \alpha^2 \cosh_{\alpha,g}(t,t_0), \quad t \in I.\end{aligned}$$
Now, by the properties of the Stieltjes–Laplace transform, we find
$$\begin{aligned}\mathscr{L}_g(f''_g(t))(z) &= \mathscr{L}_g(\alpha^2 \cos_{\alpha,g}(t,t_0))(z) \\ &= \alpha^2 \mathscr{L}_g(\cos_{\alpha,g}(t,t_0))(z) \\ &= \frac{\alpha^2 z}{z^2 - \alpha^2}.\end{aligned}$$
On the other hand, using Theorem 4.4, we obtain
$$\begin{aligned}\mathscr{L}_g(f''_g(t)) &= z^2 \mathscr{L}_g(f(t)) - zf(t_0) - f'_g(t_0) \\ &= z^2 \mathscr{L}_g(\cosh_{\alpha,g}(t,t_0))(z) - z \\ &= \frac{z^3}{z^2 - \alpha^2} - z \\ &= \frac{z^3 - z^3 + \alpha^2 z}{z^2 - \alpha^2} \\ &= \frac{\alpha^2 z}{z^2 - \alpha^2}.\end{aligned}$$

4.4 The Stieltjes–Laplace Transform of Stieltjes Integrals

In this section, we find the Stieltjes–Laplace transform for Stieltjes integrals in the case when $g \in \mathscr{C}(I)$. We have the following result.

Theorem 4.5. *Let $f \in V([s_0, \infty))$ be of exponential order λ. Then*

$$\mathscr{L}_g(F(t))(z) = \frac{1}{z}\mathscr{L}_g(f(t))(z)$$

provided that

$$\lim_{t \to \infty} F(t)e_{-z,g}(t, t_0) = 0,$$

where

$$F(t) = \int_{t_0}^{t} f(\tau)d_g\tau.$$

Proof. By definition, we have

$$\mathscr{L}_g(F(t))(z) = \int_{t_0}^{\infty} F(t)e_{-z,g}(t, t_0)d_g t$$

$$= \int_{t_0}^{\infty} F(t)e_{-z,g}(t, t_0)d_g t$$

$$= -\frac{1}{z}\int_{t_0}^{\infty} (-z)F(t)e_{-z,g}(t, t_0)d_g t$$

$$= -\frac{1}{z}\int_{t_0}^{\infty} F(t)(e_{-z,g})'_g(t)d_g t$$

$$= -\frac{1}{z}\lim_{b \to \infty}\left(F(t)e_{-z,g}(t, t_0)\Big|_{t=t_0}^{t=b}\right)$$

$$\quad + \frac{1}{z}\int_{t_0}^{\infty} f(t)e_{-z,g}(t, t_0)d_g t$$

$$= \frac{1}{z}\int_{t_0}^{\infty} f(t)e_{-z,g}(t, t_0)d_g t$$

$$= \frac{1}{z}\mathscr{L}_g(f(t))(z).$$

This completes the proof. □

Example 4.10. Let

$$f(t) = \sin_{\alpha,g}(t, t_0) - \cos_{2\alpha,g}(t, t_0),$$

$$F(t) = -\frac{1}{\alpha}\cos_{\alpha,g}(t, t_0) + \frac{1}{\alpha} - \frac{1}{2\alpha}\sin_{2\alpha,g}(t, t_0), \quad t \in I.$$

We observe that

$$F'_g(t) = f(t), \quad t \in I.$$

Then

$$\mathscr{L}_g(f(t)) = \mathscr{L}_g(\sin_{\alpha,g}(t, t_0) - \cos_{2\alpha,g}(t, t_0))$$
$$= \mathscr{L}_g(\sin_{\alpha,g}(t, t_0)) = \mathscr{L}_g(\cos_{2\alpha,g}(t, t_0))(z)$$
$$= \frac{\alpha}{z^2 + \alpha^2} - \frac{z}{z^2 + 4\alpha^2}$$
$$= \frac{\alpha z^2 + 4\alpha^3 - z^3 - \alpha^2 z}{(z^2 + \alpha^2)(z^2 + 4\alpha^2)}$$

and

$$\mathscr{L}_g(F(t))(z) = \mathscr{L}_g\left(-\frac{1}{\alpha}\cos_{\alpha,g}(t, t_0) + \frac{1}{\alpha} - \frac{1}{2\alpha}\sin_{2\alpha,g}(t, t_0)\right)$$
$$= \mathscr{L}_g\left(-\frac{1}{\alpha}\cos_{\alpha,g}(t, t_0)\right) + \mathscr{L}_g\left(\frac{1}{\alpha}\right) - \mathscr{L}_g\left(\frac{1}{2\alpha}\sin_{2\alpha,g}(t, t_0)\right)$$
$$= -\frac{1}{\alpha}\mathscr{L}_g\left(\cos_{\alpha,g}(t, t_0)\right) + \frac{1}{\alpha}\mathscr{L}_g(1) - \frac{1}{2\alpha}\mathscr{L}_g\left(\sin_{2\alpha,g}(t, t_0)\right)$$
$$= -\frac{z}{\alpha(z^2 + \alpha^2)} - \frac{1}{2\alpha} \cdot \frac{2\alpha}{z^2 + 4\alpha^2} + \frac{1}{\alpha z}$$
$$= \frac{1}{\alpha}\left(\frac{1}{z} - \frac{z}{z^2 + \alpha^2} - \frac{\alpha}{z^2 + 4\alpha^2}\right)$$
$$= \frac{1}{\alpha}\left(\frac{(z^2 + \alpha^2)(z^2 + 4\alpha^2) - z^2(z^2 + 4\alpha^2) - \alpha z(z^2 + \alpha^2)}{z(z^2 + \alpha^2)(z^2 + 4\alpha^2)}\right)$$
$$= \frac{1}{\alpha} \cdot \frac{z^4 + 4\alpha^2 z^2 + \alpha^2 z^2 + 4\alpha^4 - z^4 - 4\alpha^2 z^2 - \alpha z^3 - \alpha^3 z}{z(z^2 + \alpha^2)(z^2 + 4\alpha^2)}$$

$$= \frac{1}{\alpha} \cdot \frac{\alpha^2 z^2 + 4\alpha^4 - \alpha z^3 - \alpha^3 z}{z(z^2 + \alpha^2)(z^2 + 4\alpha^2)}$$

$$= \frac{\alpha z^2 + 4\alpha^3 - z^3 - \alpha^2 z}{z(z^2 + \alpha^2)(z^2 + 4\alpha^2)}$$

$$= \frac{1}{z}\mathscr{L}_g(f(t))(z).$$

Thus,

$$\mathscr{L}_g(F(t))(z) = \frac{1}{z}\mathscr{L}_g(f(t))(z).$$

Definition 4.3. The inverse Stieltjes–Laplace transform of a function F is a function f that has the property

$$\mathscr{L}_g(f) = F.$$

It will be denoted by $\mathscr{L}_g^{-1}(F)$.

Example 4.11. Let $I = \mathbb{R}$ and define

$$g(t) = t + 2, \quad t \in I.$$

We will find

$$\mathscr{L}_g^{-1}\left(\frac{1}{z-1}\right)(t), \quad t \in I.$$

Let $t_0 = 3$. Then, we obtain

$$\mathscr{L}_g^{-1}\left(\frac{1}{z-1}\right)(t) = e_{1,g}(t,3), \quad t \in I. \tag{4.3}$$

Example 4.12. We will find

$$\mathscr{L}_g^{-1}\left(\frac{z}{(z^2-1)(z^2+4)}\right)(t), \quad t \in I.$$

First, for $z \in \mathbb{C}$, we write

$$\frac{z}{(z^2-1)(z^2+4)} = \frac{a_1 z + a_2}{z^2-1} + \frac{a_3 z + a_4}{z^2+4},$$

where $a_1, a_2, a_3, a_4 \in \mathbb{R}$ which will be determined below. Then, we have

$$\frac{z}{(z^2-1)(z^2+4)} = \frac{(a_1 z + a_2)(z^2+4) + (a_3 z + a_4)(z^2-1)}{(z^2-1)(z^2+4)}$$

$$= \frac{a_1 z^3 + a_2 z^2 + 4a_1 z + 4a_2 + a_3 z^3 - a_3 z + a_4 z^2 - a_4}{(z^2-1)(z^2+4)}$$

$$= \frac{(a_1+a_3)z^3 + (a_2+a_4)z^2 + (4a_1-a_3)z + 4a_2 - a_4}{(z^2-1)(z^2+4)}$$

$$= \frac{(a_1+a_3)z^3 + (a_2+a_4)z^2 + (4a_1-a_2)z + 4a_2 - a_4}{(z^2-1)(z^2+4)}.$$

Thus, we get the system of equations

$$a_1 + a_3 = 0,$$
$$a_2 + a_4 = 0,$$
$$4a_1 - a_3 = 1,$$
$$4a_2 - a_4 = 0,$$

whereupon

$$a_1 = \frac{1}{5},$$
$$a_2 = 0,$$
$$a_3 = -\frac{1}{5},$$
$$a_4 = 0.$$

Hence,
$$\frac{z}{(z^2-1)(z^2+4)} = \frac{1}{5}\frac{z}{z^2-1} - \frac{1}{5}\frac{z}{z^2+4}, \quad z \in \mathbb{C}.$$

Taking the inverse Stieltjes–Laplace transform, we get

$$\mathscr{L}_g^{-1}\left(\frac{z}{(z^2-1)(z^2+4)}\right) = \mathscr{L}_g^{-1}\left(\frac{1}{5}\frac{z}{z^2-1} - \frac{1}{5}\frac{z}{z^2+4}\right)$$

$$= \mathscr{L}_g^{-1}\left(\frac{1}{5}\frac{z}{z^2-1}\right) - \mathscr{L}_g^{-1}\left(\frac{1}{5}\frac{z}{z^2+4}\right)$$

$$= \frac{1}{5}\mathscr{L}_g^{-1}\left(\frac{z}{z^2-1}\right) - \frac{1}{5}\mathscr{L}_g^{-1}\left(\frac{z}{z^2+4}\right)$$

$$= \frac{1}{5}\cosh_{1,g}(t,t_0) - \frac{1}{5}\cos_{2,g}(t,t_0).$$

Thus,
$$\mathscr{L}_g^{-1}\left(\frac{z}{(z^2-1)(z^2+4)}\right) = \frac{1}{5}\cosh_{1,g}(t,t_0) - \frac{1}{5}\cos_{2,g}(t,t_0), \quad t \in I.$$

Exercise 4.2. Find
$$\mathscr{L}_g^{-1}\left(\frac{1}{(z^2+1)^2}\right)(t), \quad t \in I.$$

Answer 4.2.
$$\frac{1}{2}\left(\sin_{1,g}(t,t_0) - \cos_{1,g}(t,t_0)\int_{t_0}^t d_g\tau - \sin_{1,g}(t,t_0)\int_{s_0}^t (g(\tau)-\tau)d_g\tau\right),$$
$t \in I.$

4.5 Advanced Practical Problems

Problem 4.1. Find $\mathscr{L}_g(f(t))(z)$, where

(1)
$$f(t) = e_{-1,g}(t,t_0) - e_{-2,g}(t,t_0), \quad t \in I,$$

(2)
$$f(t) = 2\sin_{2,g}(t,t_0) - 3\cos_{1,g}(t,t_0), \quad t \in I,$$

(3)
$$f(t) = 4e_{1,g}(t,t_0) - 4\sinh_{3,g}(t,t_0), \quad t \in I,$$

(4)
$$f(t) = \sin_{4,g}(t,t_0) - e_{3,g}(t,t_0), \quad t \in I,$$

(5)
$$f(t) = \sinh_{2,g}(t,t_0) - \cosh_{2,g}(t,t_0), \quad t \in I.$$

Answer 4.3.

(1)
$$\frac{1}{(z+1)(z+2)},$$

(2)
$$\frac{-3z^3 + 4z^2 - 12z + 4}{(z^2+1)(z^2+4)},$$

(3)
$$\frac{4(z^2 - 2z - 2)}{(z-1)(z^2-4)},$$

(4)
$$\frac{-z^2 + 4z - 28}{(z-3)(z^2+16)},$$

(5)
$$-\frac{1}{z+2}.$$

Problem 4.2. Find $\mathscr{L}_g^{-1}(f(z))(t)$, $t \in I$, where

(1)
$$g(z) = \frac{z^3 - 2z^2 + 1}{z^2(z-1)(z+2)}, \quad z \in \mathbb{C},$$

(2)
$$g(z) = \frac{2z}{z^2 + z - 20}, \quad z \in \mathbb{C},$$

(3)
$$g(z) = \frac{z}{(z^2+1)^2}, \quad z \in \mathbb{C}.$$

Answer 4.4.

(1)
$$-\frac{1}{4}t + \frac{1}{16}e_{2,g}(t,t_0) + \frac{15}{16}e_{-2,g}(t,t_0), \quad t \in I,$$

(2)
$$\frac{10}{9}e_{-5,g}(t,t_0) + \frac{8}{9}e_{4,g}(t,t_0), \quad t \in I,$$

(3)
$$\frac{1}{2}\left(-\cos_{1,g}(t,t_0) \int_{t_0}^{t} d_g\tau\right), \quad t \in I.$$

Chapter 5

First-Order Linear Stieltjes Differential Equations

This chapter deals with the investigation of homogeneous and non-homogeneous initial value problems for the first-order Stieltjes differential equations. We give the integral representations of their solutions. The Stieltjes–Laplace transform method is considered.

Throughout this chapter, we let $I \subseteq \mathbb{R}$ be such that $t_0 \in I$ and $g: I \to \mathbb{R}$ be monotone nondecreasing function that is continuous from the left everywhere. With $\mathscr{C}_g^k$, where $k \in \mathbb{N}_0$, we will denote the set of all functions $f: I \to \mathbb{R}$ for which f_g^l exist and are continuous on I for any $l \in \{0, \ldots, k\}$. For convenience, $\mathscr{C}_g^0$ we will denote with $\mathscr{C}_g$.

5.1 Linear Stieltjes Differential Operator

Definition 5.1. Suppose that $f \in \mathscr{R}_g$. We define the operator $L_1 \colon \mathscr{C}_g^1 \to \mathscr{C}_g$ by

$$L_1 x = x'_g - fx \quad \text{on } I.$$

The Stieltjes adjoint operator $L_{1g} \colon \mathscr{C}_g^1 \to \mathscr{C}_g$ is defined by

$$L_{1g} x = x'_g + fx \quad \text{on } I.$$

Theorem 5.1 (**The Stieltjes–Lagrange Identity**). *If $x, y \in \mathscr{C}_g^1$, then*

$$xL_1y + yL_{1g}x = (xy)'_g.$$

Proof. We have

$$\begin{aligned} xL_1y + yL_{1g}x &= x(y'_g - fy) + y(x'_g + fx) \\ &= xy'_g - fxy + fxy + yx'_g \\ &= xy'_g + yx'_g \\ &= (xy)'_g. \end{aligned}$$

This completes the proof. $\square$

Definition 5.2. By the solution of a given Stieltjes differential equation, we mean any function $x: I \to \mathbb{R}$ that satisfies the given Stieltjes differential equation.

Corollary 5.1. *If x and y are solutions of the equations*

$$L_1y = 0$$

and

$$L_{1g}x = 0,$$

respectively, then

$$x(t)y(t) = \text{constant} \quad \text{for } t \in I.$$

Proof. From Theorem 5.1, it follows that

$$(xy)'_g(t) = 0 \quad \text{for } t \in I.$$

Let $t_0 \in I$ be arbitrarily chosen. Integrating, in the sense of Stieltjes, the last equality from t_0 to t, we obtain

$$x(t)y(t) = x(t_0)y(t_0) \quad \text{for } t \in I.$$

This completes the proof. $\square$

5.2 Homogeneous Stieltjes Initial Value Problems

Consider the following Stieltjes initial value problem (SIVP):

$$x'_g = f(t)x, \quad t > t_0, \ t \in I, \qquad (5.1)$$
$$x(t_0) = x_0,$$

where $x_0 \in \mathbb{R}$ and $f \in \mathscr{R}_g$. We have the following result.

Theorem 5.2. *Let $x_0 \in \mathbb{R}$ and $f \in \mathscr{R}_g$. Then the unique solution of the SIVP (5.1) is given by*

$$x(t) = x_0 e_{f,g}(t, t_0) \quad \text{for } t \geq t_0. \qquad (5.2)$$

Proof. Differentiating (5.2) in the sense of Stieltjes, we obtain

$$x'_g(t) = x_0 f(t) e_{f,g}(t, t_0)$$
$$= f(t) y(t), \quad t > t_0,$$

and

$$x(t_0) = x_0 e_{f,g}(t_0, t_0)$$
$$= x_0.$$

Thus, x given by (5.2) is a solution of the IVP (5.1). Now, let y be other solution of the IVP (5.1). Then

$$\left(\frac{y}{x}\right)'_g = \frac{(y'_g(t))x(t) - y(t)x'_g(t)}{(x(t))^2}$$
$$= \frac{f(t)y(t)x(t) - f(t)y(t)x(t)}{(x(t))^2}$$
$$= 0, \quad t \geq t_0.$$

Hence, we conclude that

$$\frac{y(t)}{x(t)} = \frac{y(t_0)}{x(t_0)}$$
$$= \frac{x_0}{x_0}$$
$$= 1, \quad t \geq t_0.$$

Thus,
$$x(t) = y(t), \quad t \geq t_0.$$

Hence, the SIVP (5.1) has a unique solution given by (5.2). This completes the proof. □

Example 5.1. Let I, f, and g be as in Example 3.7 and $t_0 = 0$ and $x_0 = 1$. Then by the computations in Example 3.7 we find, for the solution of the IVP (5.1), the following representation:

$$x(t) = \begin{cases} e^{\frac{t^3}{3}+\frac{t^2}{2}} & \text{for } t \in [0,2), \\ 19e^{\frac{t^3}{3}+\frac{t^2}{2}} & \text{for } t \in [2,8]. \end{cases}$$

Example 5.2. Let I, f, and g be as in Example 3.8 and $t_0 = 0$, $x_0 = 1$. Then using the computations in Example 3.8 for the solution of the IVP (5.1), we have

$$x(3) = 60.$$

5.3 Nonhomogeneous Stieltjes Initial Value Problems

In this section, we investigate the following SIVP:

$$\begin{aligned} x'_g &= f(t)x + h(t), \quad t > t_0, \ t \in I, \\ x(t_0) &= x_0, \end{aligned} \quad (5.3)$$

where $f \in \mathscr{R}_g$ and $h \in \mathscr{C}(I)$.

Theorem 5.3. *The unique solution of the IVP* (5.3) *is given by*

$$x(t) = e_{f,g}(t,t_0)\left(x_0 + \int_{t_0}^{t} h(\tau)e_{\ominus_g f,g}(\tau,t_0)d_g\tau\right) \quad \text{for } t \geq t_0. \quad (5.4)$$

Proof. First, note that

$$e_{f,g}(t,t_0)e_{\ominus_g f,g}(t,t_0) = 1 \quad \text{for } t \geq t_0,$$

and let
$$x(t) = x_0 e_{f,g}(t,t_0) + e_{f,g}(t,t_0) \int_{t_0}^t h(\tau) e_{\ominus_g f,g}(\tau, t_0) d_g\tau \quad \text{for } t \geq t_0.$$

Then, differentiating in the sense of Stieltjes, we obtain

$$x'_g(t) = x_0 (e_{f,g})'_g(t,t_0) + \left(e_{f,g}(t,t_0) \int_{t_0}^t h(\tau) e_{\ominus f,g}(\tau, t_0) d_g\tau \right)'_g$$

$$= x_0 f(t) e_{f,g}(t,t_0) + (e_{f,g})'_g(t,t_0) \int_{t_0}^t h(\tau) e_{\ominus f,g}(\tau, t_0) d_g\tau$$

$$+ e_{f,g}(t,t_0) \left(\int_{t_0}^t h(\tau) e_{\ominus f,g}(\tau, t_0) d_g\tau \right)'_g$$

$$= x_0 f(t) e_{f,g}(t,t_0) + f(t) e_{f,g}(t,t_0) \int_{t_0}^t h(\tau) e_{\ominus f,g}(\tau, t_0) d_g\tau$$

$$+ e_{f,g}(t,t_0) h(t) e_{\ominus f,g}(t,t_0)$$

$$= f(t) \left(x_0 e_{f,g}(t,t_0) + e_{f,g}(t,t_0) \int_{t_0}^t h(\tau) e_{\ominus f,g}(\tau, t_0) d_g\tau \right) + h(t)$$

$$= f(t) x(t) + h(t),$$

i.e.,
$$x'_g(t) = f(t) x(t) + h(t), \quad t > t_0.$$

Also,
$$x(t_0) = x_0 e_{f,g}(t_0, t_0) + e_{f,g}(t_0, t_0) \int_{t_0}^{t_0} h(\tau) e_{\ominus f,g}(\tau, t_0) d_g\tau$$

$$= x_0,$$

Hence, x given by (5.4) is a solution of the SIVP (5.3). Now, assume that the SIVP (5.3) has two solutions, say x_1 and x_2. Set $y = x_1 - x_2$. Then y is a solution to the SIVP

$$x'_g = f(t) x, \quad t > t_0,$$
$$x(t_0) = 0.$$

Since the trivial solution is its solution, applying Theorem 5.2, we conclude that
$$y(t) = 0, \quad t \geq t_0,$$
i.e.,
$$x_1(t) = x_2(t), \quad t \geq t_0.$$
Hence, the SIVP (5.3) has unique solution given by (5.4). This completes the proof. □

Example 5.3. Let $I = [0, \infty)$ and $g(t) = 1 + t^2$, $t \in I$. Consider the SIVP
$$x'_g = \frac{1}{1+2t} x + \frac{-t - t^3 + 3t^2}{1+2t}, \quad t > 0,$$
$$x(0) = 1.$$

We will prove that the function,
$$x(t) = 1 + t + t^3, \quad t \in I,$$
is a solution to the considered SIVP. Actually, we have
$$x(0) = 1,$$
$$x'_g(t) = \frac{y'(t)}{g'(t)}$$
$$= \frac{1 + 3t^2}{1 + 2t}, \quad t \in I,$$

and
$$\frac{1}{1+2t} x(t) + \frac{-t - t^3 + 3t^2}{1+2t} = \frac{1 + t + t^3}{1+2t} + \frac{-t - t^3 + 3t^2}{1+2t}$$
$$= \frac{1 + 3t^2}{1+2t}$$
$$= x'_g(t), \quad t \in I.$$

5.4 The Stieltjes–Laplace Transform Method

In this section, suppose that $g \in \mathscr{C}(I)$. Consider the IVP

$$\begin{aligned} x'_g &= ax + f(t), \quad t > t_0, \quad t \in I, \\ x(t_0) &= x_0, \end{aligned} \qquad (5.5)$$

where $a, x_0 \in \mathbb{R}$ and $f \in \mathscr{C}(I)$. Taking the Stieltjes–Laplace transform of both sides of the considered equation and using the initial condition, we obtain

$$z\mathscr{L}_g(x)(z) - y_0 = a\mathscr{L}_g(x)(z) + \mathscr{L}_g(f(t))(z),$$

i.e.,

$$(z-a)\mathscr{L}_g(x)(z) = y_0 + \mathscr{L}_g(f(t))(z),$$

whereupon

$$\mathscr{L}_g(x)(z) = \frac{x_0}{z-a} + \frac{1}{z-a}\mathscr{L}_g(f(t))(z).$$

Then, taking inverse Stieltjes–Laplace transform, we obtain

$$\begin{aligned} x(t) &= \mathscr{L}_g^{-1}\left(\frac{x_0}{z-a} + \frac{1}{z-a}\mathscr{L}_g(f(t))(z)\right) \\ &= x_0 \mathscr{L}_g^{-1}\left(\frac{1}{z-a}\right) + \mathscr{L}_g^{-1}\left(\frac{1}{z-a}\mathscr{L}_g(f(t))(z)\right) \\ &= x_0 e_{a,g}(t, t_0) + \mathscr{L}_g^{-1}\left(\frac{1}{z-a}\mathscr{L}_g(f(t))(z)\right), \quad t \geq t_0. \end{aligned}$$

Thus, the solution of the SIVP (5.5) is given by

$$x(t) = x_0 e_{a,g}(t, t_0) + \mathscr{L}_g^{-1}\left(\frac{1}{z-a}\mathscr{L}_g(f(t))(z)\right), \quad t \geq t_0.$$

Example 5.4. Consider the SIVP

$$x'_g = x + e_{2,g}(t, t_0), \quad t > t_0, \quad t \in I,$$
$$x(t_0) = 1.$$

Taking the Stieltjes–Laplace transform of both sides of the first equation of the considered SIVP, we obtain

$$z\mathscr{L}_g(x)(z) - 1 = \mathscr{L}_g(x)(z) + \mathscr{L}_g(e_{2,g}(t, t_0))(z)$$
$$= \mathscr{L}_g(x)(z) + \frac{1}{z - 2},$$

i.e.,

$$(z - 1)\mathscr{L}_g(x)(z) = 1 + \frac{1}{z - 2},$$

whereupon

$$\mathscr{L}_g(x)(z) = \frac{1}{z - 1} + \frac{1}{(z - 1)(z - 2)}$$
$$= \frac{1}{z - 1} + \frac{1}{z - 2} - \frac{1}{z - 1}$$
$$= \frac{1}{z - 2}.$$

Taking the inverse Stieltjes–Laplace transform, we obtain

$$x(t) = e_{2,g}(t, t_0).$$

Hence, the solution of the considered SIVP is given by

$$x(t) = e_{2,g}(t, t_0), \quad t \geq t_0.$$

Example 5.5. Consider the SIVP

$$x'_g = x + e_{3,g}(t, t_0), \quad t > t_0,$$
$$x(t_0) = 1.$$

First-Order Linear Stieltjes Differential Equations

Taking the Stieltjes–Laplace transform of both sides of the first equation of the considered SIVP, we obtain

$$z\mathscr{L}_g(x)(z) - 1 = \mathscr{L}_g(x)(z) + \mathscr{L}_g(e_{3,g}(t,t_0))(z)$$
$$= \mathscr{L}_g(x)(z) + \frac{1}{z-3},$$

i.e.,

$$(z-1)\mathscr{L}_g(x)(z) = 1 + \frac{1}{z-3},$$

whereupon

$$\mathscr{L}_g(x)(z) = \frac{1}{z-1} + \frac{1}{(z-1)(z-3)}$$
$$= \frac{1}{z-1} + \frac{1}{2}\left(\frac{1}{z-3} - \frac{1}{z-1}\right)$$
$$= \frac{1}{2(z-1)} + \frac{1}{2(z-3)}.$$

Taking the inverse Stieltjes–Laplace transform, we obtain

$$x(t) = \frac{1}{2}(e_{1,g}(t,t_0) + e_{3,g}(t,t_0)), \quad t \geq t_0.$$

Hence, the solution of the considered SIVP is given by

$$x(t) = \frac{1}{2}(e_{1,g}(t,t_0) + e_{3,g}(t,t_0)), \quad t \geq t_0.$$

Exercise 5.1. Find the solution of the following SIVP:

$$x'_g = 2y + e_{2,g}(t,t_0) + \cos_{3,g}(t,t_0), \quad t > t_0,$$
$$x(t_0) = 1.$$

Answer 5.1.

$$X(t) = \frac{13}{10}e_{1,g}(t,t_0) + e_{2,g}(t,t_0) - \frac{3}{10}\cos_{3,g}(t,t_0) - \frac{1}{10}\sin_{3,g}(t,t_0),$$
$$t \geq t_0.$$

5.5 Advanced Practical Problems

Problem 5.1. Solve the following SIVPs:

(1) $I = [0, \infty)$, $g(t) = 1 + t^2$, $t \in I$,
$$x'_g = \sin_{1,g}(t, 0)x, \quad t > 0,$$
$$x(0) = 1.$$

(2) $I = [-1, 10]$, $g(t) = 1 + t + t^3$, $t \in I$,
$$x'_g = e_{1,g}(t, 1)x, \quad t > 1,$$
$$x(1) = 3.$$

Problem 5.2. Solve the following SIVPs:

(1) $I = (-\infty, -1]$,
$$g(t) = \begin{cases} 1 + t, & t \in (-\infty, -3], \\ 2t, & t \in (-3, -1], \end{cases}$$

$$x'_g = \cos_{2,g}(t, -5)x + e_{3,g}(t, -1), \quad t \in (-4, -1],$$
$$x(-4) = 2.$$

(2) $I = [-4, -20]$, $g(t) = -4 + 5t$, $t \in I$,
$$x'_g = 3x + \sinh_{1,g}(t, -4), \quad t \in (-4, 20],$$
$$x(-4) = 1.$$

Problem 5.3. Let $I = [0, \infty)$, $g(t) = 1 + t + t^2$, $t \in I$. Find the solution of the SIVP
$$xx'_g = 3x^2 + 1, \quad t > 0,$$
$$x(0) = 1,$$
using the change $y = x^2$.

First-Order Linear Stieltjes Differential Equations

Problem 5.4. Let $I = [0, \infty)$, $g(t) = 1 + 3t$, $t \in I$. Find the solution of the SIVP

$$\cos x (\sin x)'_g = \sin x + 1, \quad t > 0,$$
$$x(0) = -1,$$

using the change $y = \sin x$.

Problem 5.5. Let $I = [0, \infty)$, $g(t) = 1 + 4t^2$, $t \in I$. Find a solution of the SIVP

$$(1+x)^2 \left(\frac{1}{1+x}\right)'_g = \frac{1}{1+x} + e_{1,g}(t, 0), \quad t > 0,$$
$$x(0) = 2,$$

using the change $y = \frac{1}{1+x}$.

Chapter 6

Second-Order Linear Stieltjes Differential Equations

In this chapter, we investigate homogeneous and nonhomogeneous second-order linear Stieltjes differential equations. The Stieltjes–Wronskians are defined, and some of their properties are deduced. The Stieltjes analogue of the Abel theorem is proved. We define the fundamental systems of solutions, and using them, we give some representations of the general solutions of the considered classes of Stieltjes differential equations. Further, we investigate the Stieltjes–Euler–Cauchy equations. We introduce the annihilator method and the Stieltjes–Laplace transform method.

Throughout this chapter, suppose that $I \subseteq \mathbb{R}$ is such that $t_0 \in I$ and $g \colon I \to \mathbb{R}$ is a monotone nondecreasing function that is continuous on the left everywhere.

6.1 Stieltjes–Wronskians

In this section, we investigate the second-order linear Stieltjes differential equation

$$x''_g + f(t)x'_g + h(t)x = l(t), \qquad (6.1)$$

where $f, h, l \in \mathscr{C}(I)$. Define the operator $L_2 \colon \mathscr{C}^2_g(I) \to \mathscr{C}(I)$ as follows:

$$L_2 x(t) = x''_g(t) + f(t)x'_g(t) + h(t)x(t), \quad t \in I.$$

Then, (6.1) can be rewritten in the form
$$L_2 x(t) = l(t), \quad t \in I.$$

Definition 6.1. If $l \equiv 0$ on I, then (6.1) is said to be homogeneous equation. If $l \not\equiv 0$ on I, then (6.1) is said to be nonhomogeneous equation.

Theorem 6.1 (Principle of Superposition). *The operator $L_2 \colon \mathscr{C}_g^2(I) \to \mathscr{C}(I)$ is a linear operator. If x_1 and x_2 are solutions to the homogeneous equation $L_2 x = 0$, then so does $\alpha x_1 + \beta x_2$ for any $\alpha, \beta \in \mathbb{R}$.*

Proof. Let $a, b \in \mathbb{R}$ and $x, y \in \mathscr{C}_g^2(I)$ be arbitrarily chosen. Then,
$$L_2(ax+by)(t) = (ax+by)''_g(t) + f(t)(ax+by)'_g(t)$$
$$+ h(t)(ax+by)(t)$$
$$= ax''_g(t) + af(t)x'_g(t) + ah(t)x(t) + by''_g(t)$$
$$+ bf(t)y'_g(t) + bh(t)y(t)$$
$$= aL_2 x(t) + bL_2 y(t),$$

i.e.,
$$L_2(ax+by)(t) = aL_2 x(t) + bL_2 y(t), \quad t \in I.$$

Thus, $L_2 \colon \mathscr{C}_g^2(I) \to \mathscr{C}(I)$ is a linear operator. Now, using the fact
$$L_2 x_1(t) = 0, \quad t \in I$$

and
$$L_2 x_2(t) = 0, \quad t \in I,$$

we find that
$$L_2(\alpha x_1 + \beta x_2)(t) = \alpha L_2 x_1(t) + \beta L_2 x_2(t)$$
$$= 0, \quad t \in I,$$

i.e.,
$$L_2(\alpha x_1 + \beta x_2)(t) = 0, \quad t \in I.$$

This completes the proof. □

Definition 6.2. The Stieltjes differential equation (6.1) is said to be Stieltjes regressive provided

$$1 - \Delta g(t) f(t) + (\Delta g(t))^2 g(t) \neq 0, \quad t \in I.$$

Definition 6.3. For any two Stieltjes differentiable functions x_1 and x_2, we define the Stieltjes–Wronskian $W_g = W_g(x_1, x_2)$ as

$$W_g(x_1, x_2)(t) = \det \begin{pmatrix} x_1(t) & x_2(t) \\ x'_{1g}(t) & x'_{2g}(t) \end{pmatrix}, \quad t \in I.$$

We state that two solutions x_1 and x_2 of $L_2 x = 0$ form a fundamental set (or a fundamental system) of solutions of $L_2 x = 0$ provided

$$W_g(x_1, x_2)(t) \neq 0, \quad t \in I.$$

Theorem 6.2. *If the pair of functions x_1, x_2 forms a fundamental system of solutions for $L_2 x = 0$, then*

$$x(t) = a x_1(t) + b x_2(t), \quad t \in I, \tag{6.2}$$

where $a, b \in \mathbb{R}$, is a general solution of $L_2 x = 0$, i.e., every function of the form (6.2) is a solution of $L_2 x = 0$ and every solution is of the form (6.2).

Proof. Assume that the pair of functions x_1, x_2 forms a fundamental system of solutions for $L_2 x = 0$. By Theorem 6.1, it follows that any function of the form

$$a x_1(t) + b x_2(t), \quad t \in I,$$

where $a, b \in \mathbb{R}$, is a solution to $L_2 x = 0$. Now, let y be a solution of $L_2 x = 0$. Set

$$y(t_0) = y_0$$

and

$$y'_g(t_0) = y_1.$$

Denote

$$u(t) = \alpha x_1(t) + \gamma x_2(t), \quad t \in I,$$

where $\alpha, \gamma \in \mathbb{R}$ which will be determined so that
$$u(t_0) = y_0$$
and
$$u'_g(t_0) = y_1.$$
Then, we have
$$\alpha x_1(t_0) + \gamma x_2(t_0) = y_0$$
and
$$\alpha x'_{1g}(t_0) + \gamma x'_{2g}(t_0) = y_1,$$
whereupon
$$\alpha = \frac{y_0 x'_{2g}(t_0) - y_1 x_2(t_0)}{W_g(y_1, x_2)(t_0)}$$
and
$$\gamma = \frac{y_1 x_1(t_0) - y_0 x'_{1g}(t_0)}{W_g(x_1, x_2)(t_0)}.$$
Therefore,
$$u(t) = \frac{y_0 x'_{2g}(t_0) - y_1 x_2(t_0)}{W_g(y_1, x_2)(t_0)} x_1(t) + \frac{y_1 x_1(t_0) - y_0 x'_{1g}(t_0)}{W_g(x_1, x_2)(t_0)} x_2(t), \quad t \in I.$$
Now, applying the existence and uniqueness theorem (see Theorem 7.25 of Chapter 7), we get
$$u(t) = y(t), \quad t \in I.$$
This completes the proof. □

In the following, we prove some important properties of the Stieltjes–Wronskian.

Theorem 6.3. *Suppose that x_1 and x_2 are twice Stieltjes differentiable functions on I. Then, the Stieltjes–Wronskian $W_g(x_1, x_2)$ has the following properties:*

(1)
$$(W_g(x_1, x_2))'_g(t) = \det \begin{pmatrix} x_1(t) & x_2(t) \\ x''_{1g}(t) & x''_{2g}(t) \end{pmatrix}, \quad t \in I,$$

(2)
$$(W_g(x_1, x_2))'_g(t) = \det \begin{pmatrix} x_1(t) & x_2(t) \\ L_2 x_1(t) & L_2 x_2(t) \end{pmatrix}$$
$$- f(t) W_g(x_1, x_2)(t), \quad t \in I.$$

Proof. (1) We have
$$(W_g(x_1, x_2))'_g(t) = \left(\det \begin{pmatrix} x_1(t) & x_2(t) \\ x'_{1g}(t) & x'_{2g}(t) \end{pmatrix} \right)'_g$$
$$= (x_1 x'_{2g} - x_2 x'_{1g})'_g(t)$$
$$= x'_{1g}(t) x'_{2g}(t) + x_1(t) x''_{2g}(t)$$
$$- x'_{1g}(t) x'_{2g}(t) - x_2(t) x''_{1g}(t)$$
$$= x_1(t) x''_{2g}(t) - x_2(t) x''_{1g}(t)$$
$$= \det \begin{pmatrix} x_1(t) & x_2(t) \\ x''_{1g}(t) & x''_{2g}(t) \end{pmatrix}, \quad t \in I.$$

Thus,
$$(W_g(x_1, x_2))'_g(t) = \det \begin{pmatrix} x_1(t) & x_2(t) \\ x''_{1g}(t) & x''_{2g}(t) \end{pmatrix}, \quad t \in I.$$

(2) By the previous property, we get
$$(W_g(x_1, x_2))'_g(t) = \det \begin{pmatrix} x_1(t) & x_2(t) \\ x''_{1g}(t) & x''_{2g}(t) \end{pmatrix}$$
$$= \det \begin{pmatrix} x_1(t) & x_2(t) \\ L_2 x_1(t) - f(t) x'_{1g}(t) & L_2 x_2(t) - f(t) x'_{2g}(t) \\ -h(t) x_1(t) & -h(t) x_2(t) \end{pmatrix}$$

$$= \det \begin{pmatrix} x_1(t) & x_2(t) \\ L_2 x_1(t) & L_2 x_2(t) \end{pmatrix}$$

$$+ \det \begin{pmatrix} x_1(t) & x_2(t) \\ -f(t)x'_{1g}(t) - h(t)x_1(t) & -f(t)x'_{2g}(t) - h(t)x_2(t) \end{pmatrix}$$

$$= \det \begin{pmatrix} x_1(t) & x_2(t) \\ L_2 x_1(t) & L_2 x_2(t) \end{pmatrix}$$

$$+ \det \begin{pmatrix} x_1(t) & x_2(t) \\ -f(t)x'_{1g}(t) - h(t)x_1(t) & -f(t)x'_{2g}(t) - h(t)x_2(t) \end{pmatrix}$$

$$= \det \begin{pmatrix} x_1(t) & x_2(t) \\ L_2 x_1(t) & L_2 x_2(t) \end{pmatrix} + \det \begin{pmatrix} x_1(t) & x_2(t) \\ -f(t)x'_{1g}(t) & -f(t)x'_{2g}(t) \end{pmatrix}$$

$$= \det \begin{pmatrix} x_1(t) & x_2(t) \\ L_2 x_1(t) & L_2 x_2(t) \end{pmatrix} - f(t) W_g(x_1, x_2)(t), \quad t \in I.$$

Thus,

$$(W_g(x_1, x_2))'_g(t) = \det \begin{pmatrix} x_1(t) & x_2(t) \\ L_2 x_1(t) & L_2 x_2(t) \end{pmatrix} - f(t) W_g(x_1, x_2)(t),$$

$t \in I$.

This completes the proof. □

Theorem 6.4 (The Stieltjes–Abel Theorem I). *Assume that $L_2 x = 0$ is Stieltjes regressive and x_1 and x_2 are its solutions. Then,*

$$W_g(x_1, x_2)(t) = e_{-f,g}(t, s_0) W(y_1, y_2)(s_0), \quad t \in I. \quad (6.3)$$

Proof. Since x_1 and x_2 are solutions of $L_2 x = 0$, we have

$$L_2 x_1(t) = 0, \quad t \in I,$$

and

$$L_2 x_2(t) = 0, \quad t \in I.$$

This gives

$$(W_g(x_1, x_2))'_g(t) = -f(t) W_g(x_1, x_2)(t), \quad t \in I,$$

whereupon we get (6.3). This completes the proof. □

6.2 Homogeneous Second-Order Linear Stieltjes Differential Equations with Constant Coefficients

In this section, we investigate the following homogeneous second-order linear Stieltjes differential equation with constant coefficients:

$$x''_g + ax'_g + bx = 0, \tag{6.4}$$

where $a, b \in \mathbb{R}$. Suppose that (6.4) is Stieltjes regressive, i.e.,

$$1 - a\Delta g(t) + b(\Delta g(t))^2 \neq 0.$$

We will search a solution of (6.4) in the form

$$x(t) = e_{\lambda,g}(t, s_0), \quad t \in I. \tag{6.5}$$

We have

$$x'_g(t) = \lambda e_{\lambda,g}(t, s_0)$$

and

$$x''_g(t) = \lambda^2 e_{\lambda,g}(t, s_0), \quad t \in I.$$

Then,

$$x''_g(t) + ax'_g(t) + bx(t) = \lambda^2 e_{\lambda,g}(t, s_0) + a\lambda e_{\lambda,g}(t, s_0) + be_{\lambda,g}(t, s_0)$$
$$= (\lambda^2 + a\lambda + b)e_{\lambda,g}(t, s_0).$$

In view of (6.4), we obtain

$$(\lambda^2 + a\lambda + b)e_{\lambda,g}(t, s_0) = 0, \quad t \in I.$$

Since $e_{\lambda,g}(\cdot, s_0)$ does not vanish, x given in (6.5) is a solution to (6.4) if and only if

$$\lambda^2 + a\lambda + b = 0. \tag{6.6}$$

Definition 6.4. Equation (6.6) is said to be characteristic equation of the Steiltjes differential equation (6.4).

The roots of equation (6.6) are given by

$$\lambda_1 = \frac{-a - \sqrt{a^2 - 4b}}{2}$$

and
$$\lambda_2 = \frac{-a + \sqrt{a^2 - 4b}}{2}.$$

Theorem 6.5. *Equation (6.4) is Stieltjes regressive if and only if $\lambda_1, \lambda_2 \in \mathscr{R}_g$.*

Proof. Equation (6.4) is Stieltjes regressive if and only if
$$0 \neq 1 - a\Delta g(t) + b(\Delta g(t))^2$$
$$= (\lambda_1 \Delta g(t) + 1)(\lambda_2 \Delta g(t) + 1)$$

if and only if
$$1 + \lambda_1 \Delta g(t) \neq 0$$

and
$$1 + \lambda_2 \Delta g(t) \neq 0.$$

This completes the proof. □

Theorem 6.6. *Consider (6.4). Suppose that $a^2 - 4b \neq 0$. If $\Delta g(t) b - a \in \mathscr{R}_g(I)$ for $t \in I$, then*
$$e_{\lambda_1, g}(\cdot, t_0) \quad \text{and} \quad e_{\lambda_2, g}(\cdot, t_0) \qquad (6.7)$$

form a fundamental system of solutions for (6.4).

Proof. Since λ_1 and λ_2 are solutions to the characteristic equation (6.6), then (6.7) are solutions to (6.4). Next,

$$W_g(e_{\lambda_1, g}(\cdot, t_0), e_{\lambda_2, g}(\cdot, t_0))(t) = \det \begin{pmatrix} e_{\lambda_1, g}(t, t_0) & e_{\lambda_2, g}(t, t_0) \\ \lambda_1 e_{\lambda_1, g}(t, t_0) & \lambda_2 e_{\lambda_2, g}(t, t_0) \end{pmatrix}$$
$$= \lambda_2 e_{\lambda_1, g}(t, t_0) e_{\lambda_2, g}(t, s_0)$$
$$\quad - \lambda_1 e_{\lambda_1, g}(t, t_0) e_{\lambda_2, g}(t, t_0)$$
$$= (\lambda_2 - \lambda_1) e_{\lambda_1, g}(t, t_0) e_{\lambda_2, g}(t, t_0)$$
$$= \sqrt{a^2 - 4b} \, e_{\lambda_1, g}(t, t_0) e_{\lambda_2, g}(t, t_0)$$
$$\neq 0,$$

i.e.,
$$W_g(e_{\lambda_1,g}(t,t_0), e_{\lambda_2,g}(t,t_0)) \neq 0, \quad t \in I.$$
Thus, (6.7) form a fundamental system for (6.4). This completes the proof. □

Example 6.1. Let $\gamma > 0$ and $-\gamma^2 \Delta g(t) \in \mathscr{R}_g(I)$ for $t \in I$. Then
$$\cosh_{\gamma,g}(\cdot, t_0) \quad \text{and} \quad \sinh_{\gamma,g}(\cdot, t_0)$$
form a fundamental system of
$$x''_g - \gamma^2 x = 0.$$
Actually, we have
$$(\sinh_{\gamma,g})'_g(t, t_0) = \gamma \cosh_{\gamma,g}(t, t_0),$$
$$(\sinh_{\gamma,g})''_g(t) = \gamma^2 \sinh_{\gamma,g}(t, t_0),$$
$$(\cosh_{\gamma,g})'_g(t, t_0) = \gamma \sinh_{\gamma,g}(t, t_0),$$
$$(\cosh_{\gamma,g})''_g(t, t_0) = \gamma^2 \cosh_{\gamma,g}(t, t_0), \quad t \in I,$$
and
$$W_g(\sinh_{\gamma,g}(\cdot, t_0), \cosh_{\gamma,g}(\cdot, t_0))(t)$$
$$= \det \begin{pmatrix} \cosh_{\gamma,g}(t, t_0) & \sinh_{\gamma,g}(t, t_0) \\ \gamma \sinh_{\gamma,g}(t, t_0) & \gamma \cosh_{\gamma,g}(t, t_0) \end{pmatrix}$$
$$= \gamma(\cosh_{\gamma,g}(t, t_0))^2 - \gamma(\sinh_{\gamma,g}(t, t_0))^2$$
$$= \gamma$$
$$\neq 0, \quad t \in I.$$

Exercise 6.1. Suppose that $a^2 - 4b > 0$. Let
$$p = -\frac{a}{2}$$
and
$$q = \frac{\sqrt{a^2 - 4b}}{2}.$$
If $p, \Delta g(t)b - a \in \mathscr{R}_g(I)$, then prove that
$$\cosh_{\frac{q}{1+\Delta g(t)p}, g}(\cdot, s_0) e_p(\cdot, s_0) \quad \text{and} \quad \sinh_{\frac{q}{1+\Delta g(t)p}, g}(\cdot, s_0) e_p(\cdot, s_0)$$
form a fundamental system of solutions for (6.4).

6.3 Reduction of Order

In this section, we investigate the homogeneous second-order linear Stieltjes differential equation (6.4) in the case when $a^2 = 4b$. Then
$$\lambda_1 = \lambda_2 = p,$$
where $p = -\frac{a}{2}$. Hence, $a = -2p$ and $b = p^2$. Then, (6.4) takes the form
$$x''_g - 2px'_g + p^2 x = 0. \tag{6.8}$$
One solution of (6.8) is
$$x_1(t) = e_{p,g}(t, t_0), \quad t \in I.$$
We will find other linearly independent solution of (6.8). For this, we use the method of reduction of order. We will search a solution of the form
$$x(t) = v(t) e_{p,g}(t, t_0), \quad t \in I,$$
where v is Stieltjes differentiable function such that
$$v'_g(t_0) = 1, \quad v(t_0) = 0.$$
We have
$$\begin{aligned} x'_g(t) &= v'_g(t) e_{p,g}(t, t_0) + v(t) p e_{p,g}(t, t_0) \\ &= v'_g(t) e_{p,g}(t, t_0) + p v(t) e_{p,g}(t, t_0), \quad t \in I, \end{aligned}$$
and
$$\begin{aligned} x''_g(t) &= v''_g(t) e_{p,g}(t, t_0) + p v'_g(t) e_{p,g}(t, t_0) + p v'_g(t) e_{p,g}(t, t_0) \\ &\quad + p^2 v(t) e_{p,g}(t, t_0) \\ &= v''_g(t) e_{p,g}(t, t_0) + 2 p v'_g(t) e_{p,g}(t, t_0) + p^2 v(t) e_{p,g}(t, t_0), \quad t \in I. \end{aligned}$$
Now,
$$\begin{aligned} x''_g(t) - 2p x'_g(t) + p^2 x(t) &= v''_g(t) e_{p,g}(t, t_0) + 2 p v'_g(t) e_{p,g}(t, t_0) \\ &\quad + p^2 v(t) e_{p,g}(t, t_0) \\ &\quad - 2 p v'_g(t) e_{p,g}(t, t_0) - 2 p^2 v(t) e_{p,g}(t, t_0) \\ &\quad + p^2 v(t) e_{p,g}(t, t_0) \\ &= v''_g(t) e_{p,g}(t, t_0), \quad t \in I. \end{aligned}$$

In view of (6.8), we obtain
$$v_g''(t)e_{p,g}(t,t_0) = 0, \quad t \in I.$$
This gives
$$v_g'(t) = 1, \quad t \in I,$$
Now, integrating the last equation, in the sense of Stieltjes, from t_0 to t, $t \in I$, we obtain
$$v(t) = \int_{t_0}^t d_g \tau, \quad t \in I,$$
i.e.,
$$v(t) = g(t) - g(t_0), \quad t \in I.$$
Therefore, the other solution of (6.8) is
$$y(t) = (g(t) - g(t_0))e_{p,g}(t,t_0), \quad t \in I.$$

6.4 The Method of Factoring

In this section, we represent the method of factoring for a class of second-order linear Stieltjes differential equations
$$(x_g' - fx)_g'(t) - h(t)(x_g' - fx)(t) = 0, \quad t \in I, \tag{6.9}$$
where $f, h \in \mathscr{R}_g(I)$. We set
$$y(t) = x_g'(t) - f(t)x(t), \quad t \in I.$$
Then, (6.9) takes the form
$$y_g'(t) = h(t)y(t), \quad t \in I.$$
Hence
$$y(t) = ce_{h,g}(t,t_0), \quad t \in I,$$
where $c \in \mathbb{R}$. Therefore
$$x_g'(t) = f(t)x(t) + ce_{h,g}(t,t_0), \quad t \in I,$$
and
$$x(t) = e_{f,g}(t,t_0)\left(c_1 + c\int_{t_0}^t e_{\ominus f,g}(\tau,t_0)e_{h,g}(\tau,t_0)d_g\tau\right), \quad t \in I,$$
where $c_1 \in \mathbb{R}$.

Exercise 6.2. Let $I = [0, \infty)$ and define
$$g(t) = t + t^2 + 3, \quad t \in I.$$
Solve the Stieltjes differential equation
$$x''_g - 7x'_g + 12x = 0.$$

Theorem 6.7. Let p and q be constants and $f = p + q$ and $h = pq$. Then,
$$x''_g + fx'_g + hx = 0 \tag{6.10}$$
can be factored in the form (6.9).

Proof. Since $f = p + q$ and $h = pq$, we have
$$\begin{aligned} x''_g + fx'_g + hx &= x''_g + (p+q)x'_g + pqx \\ &= x''_g + px'_g + qx'_g + pqx \\ &= \left(x'_g + px\right)'_g + q\left(x'_g + px\right), \end{aligned}$$
which completes the proof. $\square$

Theorem 6.8. Let $p, q \in \mathscr{C}^1_g(I)$ and $f = -p - q$ and $h = pq - p'_g$. Then, (6.10) can be factored in the form (6.9).

Proof. Since $f = -p - q$ and $g = pq - p'_g$, we have
$$\begin{aligned} x''_g + fx'_g + hx &= x''_g - (p+q)x'_g + \left(pq - p'_g\right)x \\ &= x''_g - px'_g - qx'_g + pqx - p'_g x \\ &= \left(x'_g - px\right)'_g - q\left(x'_g - px\right), \end{aligned}$$
which completes the proof. $\square$

Example 6.2. Consider the Stieltjes differential equation
$$(x'_g - x)'_g(t) - t^2(x'_g(t) - x(t)) = 0, \quad t \in I. \tag{6.11}$$
We will prove that
$$x(t) = e_1(t, t_0), \quad t \in I,$$
is its solution. Actually, we have
$$x'_g(t) = e_1(t, t_0)$$

and
$$x'_g(t) - x(t) = 0, \quad t \in I.$$

Example 6.3. Consider the Stieltjes differential equation
$$(x'_g - x)'_g(t) + (x'_g(t) - x(t)) = 0, \quad t \in I. \tag{6.12}$$
We will prove that
$$x(t) = \cosh_{1,g}(t, t_0), \quad t \in I,$$
is its solution. We have
$$x'_g(t) = \sinh_{1,g}(t, t_0),$$
$$x'_g(t) - x(t) = \sinh_{1,g}(t, t_0) - \cosh_{1,g}(t, t_0),$$
$$(x'_g - x)'_g(t) = \cosh_{1,g}(t, t_0) - \sinh_{1,g}(t, t_0), \quad t \in I,$$
whereupon we conclude that x satisfies (6.12).

Exercise 6.3. Factor the following Stieltjes differential equations:
(1) $x''_g - (5+t)x'_g + 5tx = 0$, $t \in I$,
(2) $x''_g - (6+t)x'_g + (5+5t)x = 0$, $t \in I$.

Answer 6.1.
(1) $\left(x'_g - 5x\right)'_g - t\left(x'_g - 5x\right) = 0$, $t \in I$,
(2) $\left(x'_g - 5x\right)'_g - (t+1)\left(x'_g - 5x\right) = 0$, $t \in I$.

6.5 Stieltjes–Euler–Cauchy Equations

Suppose that $g(t) \neq 0$, $t \in I$. In this section, we investigate the second-order Stieltjes differential equation
$$(g(t))^2 x''_g + ag(t)x'_g + bx = 0, \quad t \in I, \tag{6.13}$$
where $a, b \in \mathbb{R}$.

Definition 6.5. The Stieltjes differential equation (6.13) will be called Stieltjes–Euler–Cauchy equation.

The characteristic equation of (6.13) is
$$\lambda^2 + (a-1)\lambda + b = 0. \tag{6.14}$$

Suppose that
$$(a-1)^2 - 4b > 0.$$

Then, (6.14) has two roots, say λ_1 and λ_2. Let
$$x_1(t) = e_{\frac{\lambda_1}{g(t)},g}(t,t_0), \quad t \in I,$$

and
$$x_2(t) = e_{\frac{\lambda_2}{g(t)},g}(t,t_0), \quad t \in I.$$

Then, we have
$$x'_{1g}(t) = \frac{\lambda_1}{g(t)} e_{\frac{\lambda_1}{g(t)},g}(t,t_0),$$

$$x''_{1g}(t) = -\frac{\lambda_1}{(g(t))^2} e_{\frac{\lambda_1}{g(t)},g}(t,t_0) + \frac{\lambda_1^2}{(g(t))^2} e_{\frac{\lambda_1}{g(t)},g}(t,t_0), \quad t \in I,$$

and
$$x'_{2g}(t) = \frac{\lambda_2}{g(t)} e_{\frac{\lambda_2}{g(t)},g}(t,t_0),$$

$$x''_{2g}(t) = -\frac{\lambda_2}{(g(t))^2} e_{\frac{\lambda_2}{g(t)},g}(t,t_0) + \frac{\lambda_2^2}{(g(t))^2} e_{\frac{\lambda_2}{g(t)},g}(t,t_0), \quad t \in I.$$

Hence,
$$(g(t))^2 x''_{1g}(t) + a g(t) x'_{1g}(t) + b x_1(t)$$

$$= (g(t))^2 \left(-\frac{\lambda_1}{(g(t))^2} e_{\frac{\lambda_1}{g(t)},g}(t,t_0) + \frac{\lambda_1^2}{(g(t))^2} e_{\frac{\lambda_1}{g(t)},g}(t,t_0) \right)$$

$$+ a g(t) \frac{\lambda_1}{g(t)} e_{\frac{\lambda_1}{g(t)},g}(t,t_0) + b e_{\frac{\lambda_1}{g(t)},g}(t,t_0)$$

$$= (-\lambda_1 + \lambda_1^2 + a\lambda_1 + b) e_{\frac{\lambda_1}{g(t)},g}(t,t_0)$$

$$= (\lambda_1^2 + (a-1)\lambda_1 + b) e_{\frac{\lambda_1}{g(t)},g}(t,t_0)$$

$$= 0, \quad t \in I,$$

i.e.,
$$(g(t))^2 x''_{2g}(t) + ag(t)x'_{2g}(t) + bx_2(t) = 0, \quad t \in I.$$
Thus, x_1 and x_2 are solutions to (6.13). Next,
$$W_g(x_1, x_2)(t) = \det \begin{pmatrix} x_1(t) & x_2(t) \\ x'_{1g}(t) & x'_{2g}(t) \end{pmatrix}$$
$$= \det \begin{pmatrix} x_1(t) & x_2(t) \\ \frac{\lambda_1}{g(t)} x_1(t) & \frac{\lambda_2}{g(t)} x_2(t) \end{pmatrix}$$
$$= \frac{\lambda_2}{g(t)} x_1(t) x_2(t) - \frac{\lambda_1}{g(t)} x_1(t) x_2(t)$$
$$= \frac{\lambda_2 - \lambda_1}{g(t)} x_1(t) x_2(t)$$
$$\neq 0, \quad t \in I,$$
i.e.,
$$W_g(x_1, x_2)(t) \neq 0, \quad t \in I.$$
Hence, x_1 and x_2 form a fundamental system of solutions for (6.13).

Example 6.4. Consider the Stieltjes differential equation
$$(g(t))^2 x''_g - 4g(t) x'_g + 6x = 0, \quad t \in I.$$
The corresponding characteristic equation is
$$\lambda^2 - 5\lambda + 6 = 0,$$
whose roots are $\lambda_1 = 2$ and $\lambda_2 = 3$. Therefore,
$$e_{\frac{2}{g(t)}, g}(t, t_0) \quad \text{and} \quad e_{\frac{3}{g(t)}, g}(t, t_0)$$
form a fundamental system of solutions for the considered Stieltjes differential equation.

Exercise 6.4. Find a fundamental system of solutions for the second-order Stieltjes differential equation
$$(g(t))^2 x''_g + 5g(t) x'_g + 3x = 0, \quad t \in I.$$

Answer 6.2.

$$e_{-\frac{1}{g(t)},g}(t,t_0), \quad e_{-\frac{3}{g(t)},g}(t,t_0), \quad t \in I.$$

Now, we consider the second-order Stieltjes differential equation
$$(g(t))^2 x_g'' + (1-2\alpha)g(t)x_g' + \alpha^2 x = 0, \quad t \in I, \tag{6.15}$$
where $\alpha \in \mathbb{R}$. The characteristic equation of (6.15) is
$$\lambda^2 - 2\alpha\lambda + \alpha^2 = 0.$$
Then, we see that $\lambda = \alpha$ is a root of this characteristic equation, and hence
$$x_1(t) = e_{\frac{\alpha}{g},g}(t,t_0), \quad t \in I,$$
is a solution to (6.15). Now, we will search a solution of (6.15) in the following form:
$$x_2(t) = v(t) e_{\frac{\alpha}{g},g}(t,t_0), \quad t \in I,$$
where $v \in \mathscr{C}_g^2(I)$, satisfying the following conditions:
$$x_{2g}'(t_0) = 1,$$
$$x_2(t_0) = 0.$$

We have
$$x_{2g}'(t) = v_g'(t) e_{\frac{\alpha}{g},g}(t,t_0) + \alpha \frac{v(t)}{g(t)} e_{\frac{\alpha}{g},g}(t,t_0)$$
$$= \left(v_g'(t) + \alpha \frac{v(t)}{g(t)} \right) e_{\frac{\alpha}{g},g}(t,t_0), \quad t \in I,$$

and
$$x_{2g}''(t) = v_g''(t) e_{\frac{\alpha}{g},g}(t,t_0) + \alpha \frac{v_g'(t)}{g(t)} e_{\frac{\alpha}{g},g}(t,t_0)$$
$$+ \alpha \frac{v_g'(t)g(t) - v(t)}{(g(t))^2} e_{\frac{\alpha}{g},g}(t,t_0) + \alpha^2 \frac{v(t)}{(g(t))^2} e_{\frac{\alpha}{g},g}(t,t_0)$$
$$= \frac{1}{(g(t))^2} ((g(t))^2 v_g''(t) + 2\alpha g(t) v_g'(t)$$
$$+ \alpha(\alpha-1) v(t)) e_{\frac{\alpha}{g},g}(t,t_0), \quad t \in I.$$

Hence,

$$(g(t))^2 x''_{2g}(t) + (1 - 2\alpha)x'_{2g}(t) + \alpha^2 x_2(t)$$
$$= (g(t))^2 \frac{1}{(g(t))^2}((g(t))^2 v''_g(t) + 2\alpha g(t) v'_g(t))$$
$$+ \alpha(\alpha - 1)v(t))e_{\frac{\alpha}{g},g}(t, t_0)$$
$$+ (1 - 2\alpha)g(t)\left(v'_g(t) + \alpha \frac{v(t)}{g(t)}\right) e_{\frac{\alpha}{g},g}(t, t_0) + \alpha^2 v(t) e_{\frac{\alpha}{g},g}(t, t_0)$$
$$= ((g(t))^2 v''_g(t) + (2\alpha g(t) + 1 - 2\alpha)v'_g(t)) e_{\frac{\alpha}{g},g}(t, t_0), \quad t \in I.$$

Now, in view of (6.13), we obtain

$$((g(t))^2 v''_g(t) + (2\alpha g(t) + 1 - 2\alpha)v'_g(t)) e_{\frac{\alpha}{g},g}(t, t_0) = 0.$$

Since $e_{\frac{\alpha}{g},g}(t, t_0) \neq 0$, we have

$$(g(t))^2 v''_g(t) + (2\alpha g(t) + 1 - 2\alpha)v'_g(t) = 0, \quad t \in I. \qquad (6.16)$$

Next, set

$$w(t) = v'_g(t), \quad t \in I.$$

Then

$$w(t_0) = 1,$$

and (6.16) yields

$$(g(t))^2 w'_g(t) + (2\alpha g(t) + 1 - 2\alpha)w(t) = 0, \quad t \in I.$$

Hence,

$$w'_g(t) = -\frac{2\alpha g(t) + 1 - 2\alpha}{(g(t))^2} w(t), \quad t \in I.$$

Consequently,

$$w(t) = e_{\frac{2\alpha - 1 - 2\alpha g}{g^2},g}(t, t_0), \quad t \in I,$$

and

$$v'_g(t) = e_{\frac{2\alpha - 1 - 2\alpha g}{g^2},g}(t, t_0), \quad t \in I.$$

Thus,
$$x_2(t) = \left(\int_{t_0}^{t} e_{\frac{2\alpha-1-2\alpha g}{g^2},g}(s,t_0)d_g s\right) e_{\frac{\alpha}{g},g}(t,t_0), \quad t \in I.$$

Now,
$$W_g(x_1, x_2)(t) = \det\begin{pmatrix} x_1(t) & x_2(t) \\ x'_{1g}(t) & x'_{2g}(t) \end{pmatrix}$$
$$= x_1(t)x'_{2g}(t) - x'_{1g}(t)x_2(t)$$
$$= \left(e_{\frac{\alpha}{g},g}(t,t_0)\right)^2 e_{\frac{2\alpha-1-2\alpha g}{g^2},g}(t,s_0)$$
$$- \frac{\alpha}{g(t)}\left(\int_{t_0}^{t} e_{\frac{2\alpha-1-2\alpha g}{g^2},g}(s,t_0)d_g s\right)\left(e_{\frac{\alpha}{g},g}(t,t_0)\right)^2,$$
$$t \in I.$$

Hence,
$$W_g(x_1, x_2)(t) \neq 0, \quad t \in I,$$
if and only if
$$g(t)e_{\frac{2\alpha-1-2\alpha g}{g^2},g}(t,t_0) - \alpha\left(\int_{t_0}^{t} e_{\frac{2\alpha-1-2\alpha g}{g^2},g}(s,t_0)d_g s\right) \neq 0, \quad t \in I. \tag{6.17}$$

Therefore, x_1 and x_2 form a fundamental system of solutions for (6.15) if and only if (6.17) holds.

Example 6.5. Consider the Stieltjes differential equation
$$(g(t))^2 x''_g - 9tx_g + 25x = 0, \quad t \in I.$$
The corresponding characteristic equation is
$$\lambda^2 - 10\lambda + 25 = 0.$$
We see that $\lambda = 5$ is one of its root and the functions
$$x_1(t) = e_{\frac{5}{g(t)},g}(t,t_0), \quad t \in I,$$
and
$$x_2(t) = \left(\int_{t_0}^{t} e_{\frac{9-10g}{g},g}(s,t_0)d_g s\right) e_{\frac{5}{g},g}(t,t_0), \quad t \in I,$$
form a fundamental system of solutions for the considered Stieltjes differential equation.

Exercise 6.5. Find a fundamental system of solutions for the second-order Stieltjes differential equation

$$(g(t))^2 x_g'' - 5g(t)x_g' + 9x = 0, \quad t \in I.$$

Answer 6.3.

$$e_{\frac{3}{t},g}(t,t_0), \quad t \in I,$$

and

$$\left(\int_{t_0}^t e_{\frac{5-6g}{g},g}(s,t_0)d_g s\right) e_{\frac{3}{g},g}(t,t_0), \quad t \in I.$$

6.6 Variation of Parameters

In this section, we investigate the nonhomogeneous second-order Stieltjes differential equation (6.1). Suppose that x_1 and x_2 form a fundamental system of solutions for the homogeneous second-order Stieltjes differential equation $L_2 x = 0$. We will search a particular solution of (6.1) in the form

$$x_P(t) = \gamma(t)x_1(t) + \delta(t)x_2(t), \quad t \in I,$$

where $\gamma, \delta \in \mathscr{C}_g^1(I)$ will be determined in the following. We have

$$x'_{Pg}(t) = \gamma'_g(t)x_1(t) + \gamma(t)x'_{1g}(t) + \delta'_g(t)x_2(t) + \delta(t)x'_{2g}(t), \quad t \in I.$$

Assume that

$$\gamma'_g(t)x_1(t) + \delta'_g(t)x_2(t) = 0, \quad t \in I. \tag{6.18}$$

Then,

$$x'_{Pg}(t) = \gamma(t)x'_{1g}(t) + \delta(t)x'_{2g}(t), \quad t \in I,$$

and

$$x''_{Pg}(t) = \gamma'_g(t)x'_{1g}(t) + \gamma(t)x''_{1g}(t) + \delta'_g(t)x'_{2g}(t) + \delta(t)x''_{2g}(t), \quad t \in I.$$

Hence,

$$l(t) = L_2 x_P(t)$$
$$= x''_{Pg}(t) + f(t)x'_{Pg}(t) + \delta(t)x_P(t)$$
$$= \gamma'_g(t)x'_{1g}(t) + \gamma(t)x''_{1g}(t) + \delta'_g(t)x'_{2g}(t) + \delta(t)x''_{2g}(t)$$
$$+ f(t)\gamma(t)x'_{1g}(t) + f(t)\delta(t)x'_{2g}(t) + g(t)\gamma(t)x_1(t)$$
$$+ \delta(t)\delta(t)x_2(t)$$
$$= \gamma(t)(x''_{1g}(t) + f(t)x'_{1g}(t) + \delta(t)x_1(t)) + \delta(t)(x''_{2g}(t)$$
$$+ f(t)x'_{2g}(t) + \delta(t)x_2(t))$$
$$+ \gamma'_g(t)x'_{1g}(t) + \delta'_g(t)x'_{2g}(t)$$
$$= \gamma'_g(t)x'_{1g}(t) + \delta'_g(t)x'_{2g}(t), \quad t \in I,$$

i.e.,

$$l(t) = \gamma'_g(t)x'_{1g}(t) + \delta'_g(t)x'_{2g}(t), \quad t \in I.$$

From the last equation and (6.18), we get the following system:

$$\gamma'_g(t)x_1(t) + \delta'_g(t)x_2(t) = 0,$$
$$\gamma'_g(t)x'_{1g}(t) + \delta'_g(t)x'_{2g}(t) = l(t), \quad t \in I,$$

whereupon

$$\gamma_g l'(t) = -\frac{x'_{2g}(t)l(t)}{W_g(x_1, x_2)(t)}, \quad t \in I,$$

and

$$\delta'_g(t) = \frac{x'_{1g}(t)l(t)}{W_g(x_1, x_2)(t)}, \quad t \in I.$$

Now, suppose that $\gamma(t_0) = 0$ and $\delta(t_0) = 0$. Then

$$\gamma(t) = -\int_{t_0}^{t} -\frac{y'_{2g}(\tau)l(\tau)}{W_g(x_1, x_2)(\tau)} d_g\tau \quad \text{and}$$

$$\delta(t) = \int_{t_0}^{t} \frac{x'_{1g}(\tau)h(\tau)}{W_g(x_1, x_2)(\tau)} d_g\tau, \quad t \in I.$$

Therefore, a particular solution of (6.1) is given by

$$x_P(t) = \gamma(t)x_1(t) + \delta(t)x_2(t)$$

$$= -x_1(t)\int_{t_0}^t -\frac{x'_{2g}(\tau)l(\tau)}{W_g(x_1,x_2)(\tau)}d_g\tau + x_2(t)\int_{t_0}^t \frac{x'_{1g}(\tau)l(\tau)}{W_g(x_1,x_2)(\tau)}d_g\tau$$

$$= \int_{t_0}^t \frac{x_2(t)x'_{1g}(\tau)l(\tau) - x_1(t)x'_{2g}(\tau)l(\tau)}{W_g(x_1,x_2)(\tau)}d_g\tau, \quad t \in I.$$

Hence, a general solution of (6.1) is given by

$$x(t) = c_1 x_1(t) + c_2 x_2(t)$$
$$+ \int_{t_0}^t \frac{x_2(t)x'_{1g}(\tau)l(\tau) - x_1(t)x'_{2g}(\tau)l(\tau)}{W_g(x_1,x_2)(\tau)}d_g\tau, \quad t \in I.$$

Example 6.6. We will find a general solution of the nonhomogeneous second-order Stieltjes differential equation

$$x''_g - 5x'_g + 6x = e_{4,g}(t,t_0), \quad t \in I.$$

The corresponding homogeneous equation is

$$x''_g(t) - 5x'_g(t) + 6x(t) = 0, \quad t \in I,$$

and its characteristic equation is

$$\lambda^2 - 5\lambda + 6 = 0.$$

Then the roots of this characteristic equation are $\lambda_1 = 2$ and $\lambda_2 = 3$. Hence

$$x_1(t) = e_{2,g}(t,t_0),$$
$$x_2(t) = e_{3,g}(t,t_0), \quad t \in I,$$

form a fundamental system of solutions of the corresponding homogeneous equation. We have

$$W_g(x_1, x_2)(t) = \det\begin{pmatrix} x_1(t) & x_2(t) \\ x'_{1g}(t) & x'_{2g}(t) \end{pmatrix}$$

$$= \det\begin{pmatrix} e_{2,g}(t,t_0) & e_{3,g}(t,t_0) \\ 2e_{2,g}(t,t_0) & 3e_{3,g}(t,t_0) \end{pmatrix}$$

$$= 3e_{2,g}(t,t_0)e_{3,g}(t,t_0) - 2e_{2,g}(t,t_0)e_{3,g}(t,t_0)$$

$$= e_{2,g}(t,t_0)e_{3,g}(t,t_0), \quad t \in I,$$

and

$$\int_{t_0}^{t} \frac{x'_{1g}(\tau)x_2(t) - x'_{2g}(\tau)x_1(t)}{W_g(x_1,x_2)(\tau)} l(\tau) d_g\tau$$

$$= \int_{t_0}^{t} \frac{2e_{2,g}(\tau,t_0)e_{3,g}(t,t_0) - 3e_{3,g}(\tau,t_0)e_{2,g}(t,t_0)}{e_{2,g}(\tau,t_0)e_{3,g}(\tau,t_0)} e_{4,g}(\tau,t_0) d_g\tau$$

$$= 2e_{3,g}(t,t_0)\int_{t_0}^{t} \frac{e_{4,g}(\tau,t_0)}{e_{3,g}(\tau,t_0)} d_g\tau - 3e_{2,g}(t,t_0)\int_{t_0}^{t} \frac{e_{4,g}(\tau,t_0)}{e_{2,g}(\tau,t_0)} d_g\tau$$

$$= 2e_{3,g}(t,t_0)\int_{t_0}^{t} e_{4\ominus_g 3,g}(\tau,t_0) d_g\tau - 3e_{2,g}(t,t_0)\int_{t_0}^{t} e_{4\ominus_g 2,g}(\tau,t_0) d_g\tau$$

$$= 2e_{3,g}(t,t_0)\int_{t_0}^{t} e_{\frac{1}{1+3g},g}(\tau,t_0) d_g\tau - 3e_{2,g}(t,t_0)$$

$$\int_{t_0}^{t} e_{\frac{2}{1+2g},g}(\tau,t_0) d_g\tau, \quad t \in I.$$

Thus, a general solution of the considered Stieltjes differential equation is

$$x(t) = c_1 e_{2,g}(t,t_0) + c_2 e_{3,g}(t,t_0) + 2e_{3,g}(t,t_0)\int_{t_0}^{t} e_{\frac{1}{1+3g},g}(\tau,t_0) d_g\tau$$

$$- 3e_{2,g}(t,t_0)\int_{t_0}^{t} e_{\frac{2}{1+2g},g}(\tau,t_0) d_g\tau, \quad t \in I.$$

Here, $c_1, c_2 \in \mathbb{R}$.

Exercise 6.6. Find a general solution to the second-order Stieltjes differential equation

$$(g(t))^2 x_g'' - 2g(t)x_g' + 2x = e_{\frac{3}{t},g}(t,t_0), \quad t \in I.$$

6.7 The Annihilator Method

In this section, we introduce the annihilator method.

Definition 6.6. We state that a function $f: I \to \mathbb{R}$ can be annihilated provided there is an operator of the form

$$a_0 D_g^n + a_1 D_g^{n-1} + \cdots + a_{n-1} D_g + a_n \mathscr{I},$$

where $\mathscr{I}$ is an identity operator, $D_g^k u = u_g^{(k)}$, for a function $u \in \mathscr{C}_g^k(I)$ and $a_k \in \mathbb{R}$, $k \in \{0, 1, \ldots, n\}$, so that

$$(a_0 D_g^n + a_1 D_g^{n-1} + \cdots + a_{n-1} D_g + a_n \mathscr{I})f(t) = 0, \quad t \in I.$$

Example 6.7. Let

$$f(t) = e_{2,g}(t,t_0), \quad t \in I.$$

Then

$$(D_g - 2\mathscr{I})f(t) = (D_g - 2\mathscr{I})e_{2,g}(t,t_0)$$
$$= D_g e_{2,g}(t,t_0) - 2e_{2,g}(t,t_0)$$
$$= 2e_{2,g}(t,t_0) - 2e_{2,g}(t,t_0)$$
$$= 0, \quad t \in I.$$

Thus, $D_g - 2\mathscr{I}$ is an annihilator for the function f.

Example 6.8. Consider the Stieltjes differential equation

$$x_g'' - 5x_g' + 6x = e_{5,g}(t,t_0), \quad t \in I.$$

The above equation can be rewritten in the form

$$(D_g - 2\mathscr{I})(D_g - 3\mathscr{I})y(t) = e_{5,g}(t,t_0), \quad t \in I. \tag{6.19}$$

Note that $D_g - 5\mathscr{I}$ is an annihilator for $e_{5,g}(t, t_0)$, $t \in I$. Multiplying both sides of (6.19) by $D_g - 5\mathscr{I}$, we get

$$(D_g - 5\mathscr{I})(D_g - 2\mathscr{I})(D_g - 3\mathscr{I})x(t) = 0, \quad t \in I.$$

Hence,

$$x(t) = ae_{2,g}(t, t_0) + be_{3,g}(t, t_0) + ce_{5,g}(t, t_0), \quad t \in I,$$

where $a, b, c \in \mathbb{R}$. Note that $e_{2,g}(t, t_0)$ and $e_{3,g}(t, t_0)$, $t \in I$, form a fundamental system of solutions for the homogeneous Stieltjes differential equation

$$x''_g - 5x'_g + 6x = 0, \quad t \in I.$$

Then

$$x_P(t) = ce_{5,g}(t, t_0), \quad t \in I,$$

is a particular solution to the considered Stieltjes differential equation. Thus, we have

$$e_{5,g}(t, t_0) = x''_{Pg}(t) - 5x'_{Pg}(t) + 6x_P(t)$$
$$= (25c - 5c + 6c)e_{5,g}(t, t_0)$$
$$= 26ce_{5,g}(t, t_0), \quad t \in I,$$

whereupon

$$c = \frac{1}{26}.$$

Consequently, a general solution of the considered Stieltjes differential equation is

$$x(t) = ae_{2,g}(t, t_0) + be_{3,g}(t, t_0) + \frac{1}{26}e_{5,g}(t, t_0), \quad t \in I.$$

Exercise 6.7. Use the method of annihilator to find a general solution to the second-order Stieltjes differential equation

$$x''_g + x'_g - 2x = 2 + g(t), \quad t \in I.$$

Answer 6.4.

$$x(t) = c_1 e_{1,g}(t, t_0) + c_2 e_{-2,g}(t, t_0) - \frac{5}{4} - \frac{1}{2}g(t), \quad t \in I.$$

6.8 The Stieltjes–Laplace Transform Method

Suppose that $g \in \mathscr{C}(I)$. Consider the second-order Stieltjes differential equation

$$x''_g + ax'_g + bx = f(t), \quad t \in I, \tag{6.20}$$

where $a, b \in \mathbb{R}$, $f \in \mathscr{C}(I)$, subject to the following initial conditions:

$$\begin{aligned} x(t_0) &= x_0, \\ x'_g(t_0) &= x_1, \end{aligned} \tag{6.21}$$

where $x_0, x_1 \in \mathbb{R}$. Taking the Stieltjes–Laplace transform of both sides of (6.20), we obtain, for $z \in \mathbb{C}$,

$$\begin{aligned} \mathscr{L}_g(f(t))(z) &= z^2 \mathscr{L}_g(x)(z) - zx_0 - x_1 + az\mathscr{L}_g(x)(z) - ax_0 \\ &\quad + b\mathscr{L}_g(x)(z) \\ &= (z^2 + az + b)\mathscr{L}_g(x)(z) - (a+z)x_0 - x_1, \end{aligned}$$

whereupon

$$\mathscr{L}_g(x)(z) = \frac{1}{z^2 + az + b}\mathscr{L}_g(f(t))(z) + \frac{a+z}{z^2 + az + b}x_0$$
$$+ \frac{1}{z^2 + az + b}x_1, \quad z \in \mathbb{C},$$

and hence, the solution of (6.20) subject to (6.21) is given by

$$x(t) = \mathscr{L}_g^{-1}\left(\frac{1}{z^2 + az + b}\mathscr{L}_g(f(t))(z)\right)(t)$$
$$+ \mathscr{L}_g^{-1}\left(\frac{a+z}{z^2 + az + b}\right)(t)x_0$$
$$+ \mathscr{L}_g^{-1}\left(\frac{1}{z^2 + az + b}\right)(t)x_1, \quad t \in I,$$

provided that both $\mathscr{L}_g$ and $\mathscr{L}_g^{-1}$ exist.

Example 6.9. Consider the following IVP:
$$x_g'' + 3x_g' + 2y = 1, \quad t \in I,$$
$$x(t_0) = 1,$$
$$x_g'(t_0) = 1.$$

Then its solution is given by

$$x(t) = \mathscr{L}_g^{-1}\left(\frac{1}{z^2 + 3z + 2}\mathscr{L}_g(1)(z)\right)(t) + \mathscr{L}_g^{-1}\left(\frac{z+3}{z^2 + 3z + 2}\right)(t)$$
$$+ \mathscr{L}_g^{-1}\left(\frac{1}{z^2 + 3z + 2}\right)(t)$$
$$= \mathscr{L}_g^{-1}\left(\frac{1}{z(z+1)(z+2)}\right)(t) + \mathscr{L}_g^{-1}\left(\frac{z+3}{(z+1)(z+2)}\right)(t)$$
$$+ \mathscr{L}_g^{-1}\left(\frac{1}{(z+1)(z+2)}\right)(t)$$
$$= \mathscr{L}_g^{-1}\left(\frac{1}{z(z+1)}\right)(t) - \mathscr{L}_g^{-1}\left(\frac{1}{z(z+2)}\right)(t) + \mathscr{L}_g^{-1}\left(\frac{z+3}{z+1}\right)(t)$$
$$- \mathscr{L}_g^{-1}\left(\frac{z+3}{z+2}\right)(t)$$
$$= \mathscr{L}_g^{-1}\left(\frac{1}{z}\right)(t) - \mathscr{L}_g^{-1}\left(\frac{1}{z+1}\right)(t) - \frac{1}{2}\mathscr{L}_g^{-1}\left(\frac{1}{z}\right)(t)$$
$$+ \frac{1}{2}\mathscr{L}_g^{-1}\left(\frac{1}{z+2}\right)(t) + \mathscr{L}_g^{-1}(1)(t) + 2\mathscr{L}_g^{-1}\left(\frac{1}{z+1}\right)(t)$$
$$- \mathscr{L}_g^{-1}(1)(t) - \mathscr{L}_g^{-1}\left(\frac{1}{z+2}\right)(t)$$
$$= 1 - e_{-1,g}(t, t_0) - \frac{1}{2} + \frac{1}{2}e_{2,g}(t, t_0) + 2e_{-1,g}(t, t_0) - e_{-2,g}(t, t_0)$$
$$= \frac{1}{2} + \frac{1}{2}e_{-2,g}(t, t_0) + e_{-1,g}(t, t_0),$$

i.e.,
$$x(t) = \frac{1}{2} + \frac{1}{2}e_{-2,g}(t, t_0) + e_{-1,g}(t, t_0), \quad t \in I.$$

Exercise 6.8. Using the Stieltjes–Laplace transform method, solve the following IVP:

$$x''_g + x'_g - 2x = 0, \quad t \in I,$$
$$x(t_0) = 1,$$
$$x'_g(t_0) = 1.$$

Answer 6.5.

$$x(t) = e_{1,g}(t, t_0), \quad t \in I.$$

6.9 Advanced Practical Problems

Problem 6.1. Suppose that $a^2 - 4b < 0$. Let

$$p = -\frac{a}{2}$$

and

$$q = \frac{\sqrt{4b - a^2}}{2}.$$

If $p, \Delta g(t)b - a \in \mathcal{R}_g(I)$, then prove that

$$\cos\tfrac{q}{1+\Delta g(t)p}, g(\cdot, t_0) e_{p,g}(\cdot, t_0), \quad \sin\tfrac{q}{1+\Delta g(t)p}, g(\cdot, t_0) e_{p,g}(\cdot, t_0)$$

form a fundamental system of solutions for the homogeneous Stieltjes differential equation (6.4).

Problem 6.2. Let $I = \mathbb{R}$ and define

$$g(t) = t^3 + 3t + 1, \quad t \in I.$$

Then solve the Stieltjes differential equation

$$x''_g - 5x'_g + 4x = 0, \quad t \in I.$$

Problem 6.3. Factorize the following Stieltjes differential equations:
(1) $x''_g - 6x'_g + 5x = 0, \; t \in I,$
(2) $x''_g - 8x'_g + 7x = 0, \; t \in I.$

Answer 6.6.

(1) $D_g(D_g x - 5x) - (D_g x - 5x) = 0$, $t \in I$,
(2) $D_g(D_g x - 7x) - (D_g x - 7x) = 0$, $t \in I$.

Problem 6.4. Find a fundamental system of solutions for the Stieltjes differential equation

$$(g(t))^2 x_g'' - 5t x_g' + 8x = 0, \quad t \in I.$$

Problem 6.5. Find a fundamental system of solutions for the Stieltjes differential equation

$$(g(t))^2 x_g'' - 7g(t) x_g' + 16x = 0, \quad t \in I.$$

Problem 6.6. Find a general solution to the Stieltjes differential equation

$$x_g'' - 3x_g' + 2x = e_{2,g}(t, t_0), \quad t \in I.$$

Problem 6.7. Use the method of annihilator to find a general solution to the Stieltjes differential equation

$$x_g'' + x = e_{3,g}(t, t_0), \quad t \in I.$$

Answer 6.7.

$$x(t) = c_1 \sin_{1,g}(t, t_0) + c_2 \cos_{1,g}(t, t_0) + \frac{1}{10} e_{3,g}(t, t_0), \quad t \in I.$$

Problem 6.8. Using the Stieltjes–Laplace transform method, solve the following SIVP:

$$x_g'' - 9x = 0, \quad t \in I,$$
$$x(t_0) = 0,$$
$$x_g'(t_0) = 1.$$

Answer 6.8.

$$x(t) = \frac{1}{3} \sinh_{3,g}(t, t_0), \quad t \in I.$$

Problem 6.9. Using the Stieltjes–Laplace transform method, solve the following SIVP:

$$x_g'' - 10x_g' + 9x = 0, \quad t \in I,$$
$$x(y_0) = 1,$$
$$x_g'(t_0) = -1.$$

Problem 6.10. Using the Stieltjes–Laplace transform method, solve the following SIVP:

$$x_g'' - 16x_g' + 64x = 0, \quad t \in I,$$
$$x(t_0) = -1,$$
$$x_g'(t_0) = 3.$$

Chapter 7

Stieltjes Differential Systems

In this chapter, we introduce the Stieltjes differential systems and investigate their structures. We define the Stieltjes exponential matrix-valued function and deduct some of its properties. The constant case is also considered. We prove the Stieltjes analogue of the classical Liouville theorem. The Stieltjes–Putzer algorithm for finding the Stieltjes exponential matrix-valued function in the case when the matrix is constant will be presented.

Suppose that $I \subseteq \mathbb{R}$ is such that $t_0 \in I$ and $g \colon I \to \mathbb{R}$ is a monotone nondecreasing function that is continuous from the left everywhere.

7.1 Structure of Stieltjes Differential Systems

Suppose that A is an $m \times n$-matrix on I, $A = (a_{ij})_{1 \leq i \leq m, 1 \leq j \leq n}$, in short $A = (a_{ij})$, $a_{ij} \colon I \to \mathbb{R}$, $1 \leq i \leq m$, $1 \leq j \leq n$. A is also known as an $m \times n$-matrix-valued function defined on I. The $m \times n$-identity matrix will denoted by $\mathscr{I}$ and the $m \times n$-zero matrix by O.

Definition 7.1. We state that $m \times n$-matrix A is Stieltjes differentiable on I provided each entry of A is Stieltjes differentiable on I. In this case, we write

$$A'_g = (a'_{ijg}).$$

Example 7.1. Let $I = [0, \infty)$. Define $g(t) = t^2 + t + 1$, $t \in I$, and
$$A(t) = \begin{pmatrix} t+1 & t^2+t \\ 2t-3 & 2t^2-3t+2 \end{pmatrix}, \quad t \in I.$$
We will find $A'_g(t)$, $t \in I$. We have
$$a_{11}(t) = t+1,$$
$$a_{12}(t) = t^2+t,$$
$$a_{21}(t) = 2t-3,$$
$$a_{22}(t) = 2t^2-3t+2, \quad t \in I.$$
Note that $g \in \mathscr{C}^1(I)$ and $a_{ij} \in \mathscr{C}^1(I)$, $i, j \in \{1, 2\}$. Then,
$$a'_{11}(t) = 1,$$
$$a'_{12}(t) = 2t+1,$$
$$a'_{21}(t) = 2,$$
$$a'_{22}(t) = 4t-3,$$
$$g'(t) = 2t+1 \quad t \in I.$$
Therefore,
$$a'_{11g}(t) = \frac{a'_{11}(t)}{g'(t)}$$
$$= \frac{1}{2t+1},$$
$$a'_{12g}(t) = \frac{a'_{12}(t)}{g'(t)}$$
$$= \frac{2t+1}{2t+1}$$
$$= 1,$$
$$a'_{21g}(t) = \frac{a'_{21}(t)}{g'(t)}$$
$$= \frac{2}{2t+1},$$

$$a'_{22g}(t) = \frac{a'_{22}(t)}{g'(t)}$$
$$= \frac{4t-3}{2t+1}, \quad t \in I,$$

and

$$A'_g(t) = \begin{pmatrix} a'_{11g}(t) & a'_{12g}(t) \\ a'_{21g}(t) & a'_{22g}(t) \end{pmatrix}$$
$$= \begin{pmatrix} \frac{1}{2t+1} & 1 \\ \frac{2}{2t+1} & \frac{4t-3}{2t+1} \end{pmatrix}, \quad t \in I.$$

Example 7.2. Let $I = [0, \infty)$ and

$$a_{11}(t) = \begin{cases} -3 & \text{for } t \in [0,1], \\ t+4 & \text{for } t \in (1, \infty), \end{cases}$$

$$a_{12}(t) = \begin{cases} t+1 & \text{for } t \in [0,1], \\ t^2 & \text{for } t \in (1, \infty), \end{cases}$$

$$a_{21}(t) = \begin{cases} -t & \text{for } t \in [0,1], \\ t+7 & \text{for } t \in (1, \infty), \end{cases}$$

$$a_{22}(t) = \begin{cases} t-1 & \text{for } t \in [0,1], \\ 8 & \text{for } t \in (1, \infty), \end{cases}$$

$$g(t) = \begin{cases} 1 & \text{for } t \in [0,1], \\ 6 & \text{for } t \in (1, \infty), \end{cases}$$

$$A(t) = \begin{pmatrix} a_{11}(t) & a_{12}(t) \\ a_{21}(t) & a_{22}(t) \end{pmatrix}, \quad t \in I.$$

We will find $A'_g(1)$. We have

$$a_{11}(1+) = 1+4$$
$$= 5,$$
$$a_{11}(1) = -3,$$

$$a_{12}(1+) = 1^2$$
$$= 1,$$
$$a_{12}(1) = 1+1$$
$$= 2,$$
$$a_{21}(1+) = 1+7$$
$$= 8,$$
$$a_{21}(1) = -1,$$
$$a_{22}(1+) = 8,$$
$$a_{22}(1) = 1-1$$
$$= 0,$$
$$g(1+) = 6,$$
$$g(1) = 1.$$

Hence,

$$a'_{11g}(1) = \frac{a_{11}(1+) - a_{11}(1)}{g(1+) - g(1)}$$
$$= \frac{5-3}{6-1}$$
$$= \frac{2}{5},$$
$$a'_{12g}(1) = \frac{a_{12}(1+) - a_{12}(1)}{g(1+) - g(1)}$$
$$= \frac{1-2}{6-1}$$
$$= -\frac{1}{5},$$
$$a'_{21g}(1) = \frac{a_{21}(1+) - a_{21}(1)}{g(1+) - g(1)}$$
$$= \frac{8-(-1)}{6-1}$$

$$= \frac{9}{5},$$

$$a'_{22g}(1) = \frac{a_{22}(1+) - a_{22}(1)}{g(1+) - g(1)}$$

$$= \frac{8 - 0}{6 - 1}$$

$$= \frac{8}{5},$$

and

$$A'_g(t) = \begin{pmatrix} a'_{11g}(t) & a'_{12g}(t) \\ a'_{21g}(t) & a'_{22g}(t) \end{pmatrix}$$

$$= \begin{pmatrix} \dfrac{2}{5} & -\dfrac{1}{5} \\ \dfrac{9}{5} & \dfrac{8}{5} \end{pmatrix}.$$

Example 7.3. Let $I = [0, \infty)$ and

$$a_{11}(t) = \sin t + 2,$$

$$a_{12}(t) = \begin{cases} t - 2 & \text{for } t \in [0, 2], \\ 4 & \text{for } t \in (2, \infty), \end{cases}$$

$$a_{13}(t) = t^2,$$

$$a_{21}(t) = 1 + 3t,$$

$$a_{22}(t) = \begin{cases} t^2 & \text{for } t \in [0, 2], \\ 7 & \text{for } t \in (2, \infty), \end{cases}$$

$$a_{23}(t) = \begin{cases} -3 & \text{for } t \in [0, 2], \\ 4 & \text{for } t \in (2, \infty), \end{cases}$$

$$a_{31}(t) = \begin{cases} 1 & \text{for } t \in [0,2], \\ t+2 & \text{for } t \in (2,\infty), \end{cases}$$

$$a_{32}(t) = 1,$$

$$a_{33}(t) = \begin{cases} 0 & \text{for } t \in [0,2], \\ 6 & \text{for } t \in (2,\infty), \end{cases}$$

$$g(t) = \begin{cases} t & \text{for } t \in [0,2], \\ 2-8t & \text{for } t \in (2,\infty), \end{cases}$$

$$A(t) = \begin{pmatrix} a_{11}(t) & a_{12}(t) & a_{13}(t) \\ a_{21}(t) & a_{22}(t) & a_{23}(t) \\ a_{31}(t) & a_{32}(t) & a_{33}(t) \end{pmatrix}, \quad t \in I.$$

We will find $A'_g(2)$. We have

$$a_{11}(2+) = \sin 2 + 2,$$
$$a_{11}(2) = \sin 2 + 2,$$
$$a_{12}(2+) = 4,$$
$$a_{12}(2) = 2 - 2$$
$$= 0,$$
$$a_{13}(2+) = 2^2$$
$$= 4,$$
$$a_{13}(2) = 4,$$
$$a_{21}(2+) = 1 + 3 \cdot 2$$
$$= 7,$$
$$a_{21}(2) = 7,$$
$$a_{22}(2+) = 7,$$
$$a_{22}(2) = 2^2$$
$$= 4,$$
$$a_{23}(2+) = 4,$$

$$a_{23}(2) = -3,$$
$$a_{31}(2+) = 2+2$$
$$= 4,$$
$$a_{31}(2) = 1,$$
$$a_{32}(2+) = 1,$$
$$a_{32}(2) = 1,$$
$$a_{33}(2+) = 6,$$
$$a_{33}(2) = 0,$$
$$g(2+) = 2 - 8 \cdot 2$$
$$= -14,$$
$$g(2) = 2.$$

Hence,

$$a'_{11g}(2) = \frac{a_{11}(2+0) - a_{11}(2)}{g(2+0) - g(2)}$$
$$= \frac{\sin 2 + 2 - \sin 2 - 2}{2 + 14}$$
$$= 0,$$
$$a'_{12g}(2) = \frac{a_{12}(2+0) - a_{12}(2)}{g(2+0) - g(2)}$$
$$= \frac{4 - 0}{2 + 14}$$
$$= \frac{1}{8},$$
$$a'_{13g}(2) = \frac{a_{13}(2+0) - a_{13}(2)}{g(2+0) - g(2)}$$
$$= \frac{4 - 4}{2 + 14}$$
$$= 0,$$
$$a'_{21g}(2) = \frac{a_{21}(2+0) - a_{21}(2)}{g(2+0) - g(2)}$$

$$= \frac{7-7}{2+14}$$
$$= 0,$$

$$a'_{22g}(2) = \frac{a_{22}(2+0) - a_{22}(2)}{g(2+0) - g(2)}$$
$$= \frac{7-4}{2+14}$$
$$= \frac{3}{16},$$

$$a'_{23g}(2) = \frac{a_{23}(2+0) - a_{23}(2)}{g(2+0) - g(2)}$$
$$= \frac{4+3}{2+14}$$
$$= \frac{7}{16},$$

$$a'_{31g}(2) = \frac{a_{31}(2+0) - a_{31}(2)}{g(2+0) - g(2)}$$
$$= \frac{1-1}{2+14}$$
$$= 0,$$

$$a'_{32g}(2) = \frac{a_{32}(2+0) - a_{32}(2)}{g(2+0) - g(2)}$$
$$= \frac{1-1}{2+14}$$
$$= 0,$$

$$a'_{33g}(2) = \frac{a_{33}(2+0) - a_{33}(2)}{g(2+0) - g(2)}$$
$$= \frac{0-6}{2+14}$$
$$= -\frac{3}{8},$$

and
$$A'_g(2) = \begin{pmatrix} a'_{11g}(2) & a'_{12g}(2) & a'_{13g}(2) \\ a'_{21g}(2) & a'_{22g}(2) & a'_{23g}(2) \\ a'_{31g}(2)m & a'_{32g}(2) & a'_{33g}(2) \end{pmatrix}$$

$$= \begin{pmatrix} 0 & \frac{1}{8} & 0 \\ 0 & \frac{3}{16} & \frac{7}{16} \\ 0 & 0 & -\frac{3}{8} \end{pmatrix}.$$

Exercise 7.1. Let $I = \mathbb{R}$. Define $g(t) = t^3 + t + 3$, $t \in I$, and
$$A(t) = \begin{pmatrix} t^3 & t^2 \\ 2t+4 & t-1 \end{pmatrix}, \quad t \in I.$$

Find $A'_g(t)$, $t \in I$.

Answer 7.1.
$$A'_g(t) = \begin{pmatrix} \dfrac{3t^2}{3t^2+1} & \dfrac{2t}{3t^2+1} \\ \dfrac{2}{3t^2+1} & \dfrac{1}{3t^2+1} \end{pmatrix}, \quad t \in I.$$

In the following, we suppose that $B = (b_{ij})_{1 \le i \le m, 1 \le j \le n}$, $b_{ij}: I \to \mathbb{R}$, $1 \le i \le m$, $1 \le j \le n$.

Theorem 7.1. *Let A and B be Stieltjes differentiable $m \times n$-matrix-valued functions on I. Then*
$$(A+B)'_g = A'_g + B'_g \quad \text{on } I.$$

Proof. We have
$$(A+B)(t) = (a_{ij}(t) + b_{ij}(t)),$$
$$(A+B)'_g(t) = \left(a'_{ijg}(t) + b'_{ijg}(t)\right)$$
$$= \left(a'_{ijg}(t)\right) + \left(b'_{ijg}(t)\right)$$
$$= A'_g(t) + B'_g(t), \quad t \in I.$$

This completes the proof. □

Theorem 7.2. Let $\alpha \in \mathbb{R}$ and A be a Stieltjes differentiable $m \times n$-matrix-valued function on I. Then

$$(\alpha A)'_g = \alpha A'_g \quad \text{on } I.$$

Proof. We have

$$(\alpha A)'_g(t) = \left((\alpha a_{ij})'_g(t)\right)$$
$$= (\alpha a'_{ijg}(t))$$
$$= \alpha \left(a'_{ijg}(t)\right)$$
$$= \alpha A'_g(t), \quad t \in I.$$

This completes the proof. □

Theorem 7.3. Let A and B be two Stieltjes differentiable $n \times n$-matrix-valued functions on I. Then

$$(AB)'_g(t) = A'_g(t)B(t+) + A(t+)B'_g(t), \quad t \in I.$$

Proof. We have

$$(AB)(t) = \left(\sum_{k=1}^{n} a_{ik}(t)b_{kj}(t)\right), \quad t \in I.$$

Then,

$$(AB)'_g(t) = \left(\left(\sum_{k=1}^{n} a_{ik}b_{kj}\right)'_g(t)\right)$$
$$= \left(\sum_{k=1}^{n}(a_{ik}b_{kj})'_g(t)\right)$$
$$= \left(\sum_{k=1}^{n}\left(a'_{ikg}(t)b_{kj}(t+) + a_{ik}(t+)b'_{kjg}(t)\right)\right)$$
$$= \left(\sum_{k=1}^{n} a'_{ikg}(t)b_{kj}(t+)\right) + \left(\sum_{k=1}^{n} a_{ik}(t+)b'_{kjg}(t)\right)$$
$$= A'_g(t)B(t+) + A(t+)B'_g(t), \quad t \in I.$$

Thus,
$$(AB)'_g(t) = A'_g(t)B(t+) + A(t+)B'_g(t), \quad t \in I.$$
This completes the proof. □

Example 7.4. Let $I = [0, \infty)$. Define $g(t) = \frac{1}{2}t$,
$$A(t) = \begin{pmatrix} t & t-1 \\ 2 & 3t+1 \end{pmatrix}, \quad \text{and} \quad B(t) = \begin{pmatrix} 1 & t \\ t+1 & t-1 \end{pmatrix}, \quad t \in I.$$

Then
$$(AB)(t) = \begin{pmatrix} t & t-1 \\ 2 & 3t+1 \end{pmatrix} \begin{pmatrix} 1 & t \\ t+1 & t-1 \end{pmatrix}$$
$$= \begin{pmatrix} t^2+t-1 & 2t^2-2t+1 \\ 3t^2+4t+3 & 3t^2-1 \end{pmatrix}$$
$$= C(t)$$
$$= (c_{ij}(t)), \quad t \in I.$$

We have
$$a_{11}(t) = t,$$
$$a_{12}(t) = t-1,$$
$$a_{21}(t) = 2,$$
$$a_{22}(t) = 3t+1,$$
$$b_{11}(t) = 1,$$
$$b_{12}(t) = t,$$
$$b_{21}(t) = t+1,$$
$$b_{22}(t) = t-1,$$
$$c_{11}(t) = t^2+t-1,$$
$$c_{12}(t) = 2t^2-2t+1,$$
$$c_{21}(t) = 3t^2+4t+3,$$
$$c_{22}(t) = 3t^2-1, \quad t \in I.$$

Then

$$a'_{11}(t) = 1,$$
$$a'_{12}(t) = 1,$$
$$a'_{21}(t) = 0,$$
$$a'_{22}(t) = 3,$$
$$b'_{11}(t) = 0,$$
$$b'_{12}(t) = 1,$$
$$b'_{21}(t) = 1,$$
$$b'_{22}(t)(t) = 1,$$
$$c'_{11}(t) = 2t + 1,$$
$$c'_{12}t(t) = 4t - 2,$$
$$c'_{21}(t) = 6t + 4,$$
$$c'_{22}(t) = 6t$$
$$g'(t) = \frac{1}{2}, \quad t \in I.$$

Hence

$$a'_{11g}(t) = 1,$$
$$a'_{12g}(t) = 1,$$
$$a'_{21g}(t) = 0,$$
$$a'_{22g}(t) = 3,$$
$$b'_{11g}(t) = 0,$$
$$b'_{12g}(t) = 1,$$
$$b'_{21g}(t) = 1,$$
$$b'_{22g}(t)(t) = 1,$$
$$c'_{11g}(t) = 2t + 1,$$
$$c'_{12g}t(t) = 4t - 2,$$

$$c'_{21g}(t) = 6t + 4,$$
$$c'_{22g}(t) = 6t, \quad t \in I.$$

Therefore
$$(AB)'_g(t) = \begin{pmatrix} 4t+2 & 8t-4 \\ 12t+8 & 12t \end{pmatrix}, \quad t \in I.$$

Also,
$$A'_g(t)B(t) = \begin{pmatrix} 2 & 2 \\ 0 & 6 \end{pmatrix} \begin{pmatrix} 1 & t \\ t+1 & t-1 \end{pmatrix}$$
$$= \begin{pmatrix} 2t+4 & 4t-2 \\ 6t+6 & 6t-6 \end{pmatrix},$$

$$A(t)B'_g(t) = \begin{pmatrix} t & t-1 \\ 2 & 3t+1 \end{pmatrix} \begin{pmatrix} 0 & 2 \\ 2 & 2 \end{pmatrix}$$
$$= \begin{pmatrix} 2t-2 & 4t-2 \\ 6t+2 & 6t+6 \end{pmatrix},$$

$$A'_g(t)B(t) + A(t)B'_g(t) = \begin{pmatrix} 2t+4 & 4t-2 \\ 6t+6 & 6t-6 \end{pmatrix} + \begin{pmatrix} 2t-2 & 4t-2 \\ 6t+2 & 6t+6 \end{pmatrix}$$
$$= \begin{pmatrix} 4t+2 & 8t-4 \\ 12t+8 & 12t \end{pmatrix}, \quad t \in I.$$

Consequently,
$$(AB)'_g(t) = A'_g(t)B(t) + A(t)B'_g(t), \quad t \in I.$$

Exercise 7.2. Let $I = [0, \infty)$ and define $g(t) = \frac{1}{5}t + 7$,
$$A(t) = \begin{pmatrix} t^2+1 & t-2 \\ 2t-1 & t+1 \end{pmatrix},$$
and
$$B(t) = \begin{pmatrix} t & 2t+1 \\ t & t-1 \end{pmatrix}, \quad t \in I.$$

Then prove that
$$(AB)'_g(t) = A'_g(t)B(t) + A(t)B'_g(t), \quad t \in I.$$

Theorem 7.4. *Let A be a Stieltjes differentiable $n \times n$-matrix-valued function on I and A^{-1} exists on I. Then*
$$\left(A^{-1}\right)'_g(t) = -A^{-1}(t+)A'_g(t)\left(A\right)^{-1}(t+), \quad t \in I.$$

Proof. We have
$$\mathscr{I} = AA^{-1} \quad \text{on } I,$$
and
$$\mathscr{I}'_g(t) = 0, \quad t \in I,$$
whereupon, using Theorem 7.3, we get
$$\mathscr{I}'_g(t) = (AA^{-1})'_g(t)$$
$$= A'_g(t)A^{-1}(t+) + A(t+)(A^{-1})'_g(t), \quad t \in I,$$
whereupon
$$A(t+)(A^{-1})'_g(t) = -A'_g(t)A^{-1}(t+), \quad t \in I,$$
and hence,
$$(A^{-1})'_g(t) = -A^{-1}(t+)A'_g(t)A^{-1}(t+), \quad t \in I.$$
This completes the proof. □

Exercise 7.3. Let A and B be Stieltjes differentiable $n \times n$-matrix-valued functions on I such that B^{-1} exists on I. Then prove that
$$(AB^{-1})'_g = A'_g B^{-1} - AB^{-1}B'_g B^{-1} \quad \text{on } I.$$

Definition 7.2. We state that a matrix-valued function A is continuous on I provided each entry of A is continuous. The class of such continuous $m \times n$-matrix-valued functions on I is denoted by
$$\mathscr{C} = \mathscr{C}(I) = \mathscr{C}(I, \mathbb{R}^{m \times n}).$$

In the following, we suppose that A and B are $n \times n$-matrix-valued functions.

Definition 7.3. We state that an $n \times n$-matrix-valued function A on I is Stieltjes regressive with respect to I provided

$$\mathscr{I} + \Delta g(t) A(t) \quad \text{is invertible for all } t \in I.$$

The class of such Stieltjes regressive and continuous functions is denoted, similar to the scalar case, by

$$\mathscr{R}_g = \mathscr{R}_g(I) = \mathscr{R}_g(I, \mathbb{R}^{n \times n}).$$

Theorem 7.5. *The $n \times n$-matrix-valued function A is Stieltjes regressive if and only if the eigenvalues $\lambda_i(t)$ of $A(t)$ are Stieltjes regressive for all $1 \leq i \leq n$.*

Proof. Let $i \in \{1, \ldots, n\}$ be arbitrarily chosen and $\lambda_i(t)$ be an eigenvalue corresponding to the eigenvector $y(t)$. Then

$$\begin{aligned}(1 + \Delta g(t) \lambda_i(t)) \, x(t) &= \mathscr{I} x(t) + \Delta g(t) \lambda_i(t) x(t) \\ &= \mathscr{I} x(t) + \Delta g(t) A(t) x(t) \\ &= (\mathscr{I} + \Delta g(t) A(t)) \, x(t),\end{aligned}$$

whereupon it follows the assertion. This completes the proof. □

Example 7.5. Let $I = 3\mathbb{Z}$. Define $g(t) = \frac{t}{2} + 3$ and

$$A(t) = \begin{pmatrix} 5t & t \\ 4t & 2t \end{pmatrix}, \quad t \in I.$$

Consider the equation

$$\det \begin{pmatrix} 5t - \lambda(t) & t \\ 4t & 2t - \lambda(t) \end{pmatrix} = 0, \quad t \in I.$$

The corresponding characteristic equation is

$$(\lambda(t) - 5t)(\lambda(t) - 2t) - 4t^2 = 0, \quad t \in I,$$

i.e.,

$$(\lambda(t))^2 - 7t\lambda(t) + 10t^2 - 4t^2 = 0, \quad t \in I,$$

i.e.,
$$(\lambda(t))^2 - 7t\lambda(t) + 6t^2 = 0, \quad t \in I.$$

The roots of this characteristic equation are given by
$$\lambda_{1,2}(t) = \frac{7t \pm \sqrt{49t^2 - 24t^2}}{2}$$
$$= \frac{7t \pm 5t}{2}, \quad t \in I,$$

i.e.,
$$\lambda_1(t) = 6t,$$
$$\lambda_2(t) = t, \quad t \in I.$$

Now, using the fact that $\Delta g(t) = 0$, $t \in I$, we get
$$1 + \Delta g(t)\lambda_{1,2}(t) \neq 0, t \in I.$$

Consequently, the matrix A is Stieltjes regressive.

Theorem 7.6. *Let A be a 2×2-matrix-valued function. Then, A is Stieltjes regressive if and only if*
$$\operatorname{tr} A + \Delta g(t) \det A, t \in I,$$
is Stieltjes regressive. Here, $\operatorname{tr} A$ denotes the trace of the matrix A.

Proof. Let
$$A(t) = \begin{pmatrix} a_{11}(t) & a_{12}(t) \\ a_{21}(t) & a_{22}(t) \end{pmatrix}, \quad t \in I.$$

Then
$$\mathscr{I} + \Delta g(t)A(t) = \begin{pmatrix} 1 & 0 \\ 0 & 1 \end{pmatrix} + \begin{pmatrix} \Delta g(t)a_{11}(t) & \Delta g(t)a_{12}(t) \\ \Delta g(t)a_{21}(t) & \Delta g(t)a_{22}(t) \end{pmatrix}$$
$$= \begin{pmatrix} 1 + \Delta g(t)a_{11}(t) & \Delta g(t)a_{12}(t) \\ \Delta g(t)a_{21}(t) & 1 + \Delta g(t)a_{22}(t) \end{pmatrix}, \quad t \in I.$$

Therefore, we obtain

$$\det(\mathscr{I} + \Delta g(t)A(t))$$
$$= (1 + \Delta g(t)a_{11}(t))(1 + \Delta g(t)a_{22}(t))$$
$$- (\Delta g(t))^2 a_{12}(t)a_{21}(t)$$
$$= 1 + \Delta g(t)a_{22}(t) + \Delta g(t)a_{11}(t)$$
$$+ (\Delta g(t))^2 a_{11}(t)a_{22}(t) - (\Delta g(t))^2 a_{12}(t)a_{21}(t)$$
$$= 1 + \Delta g(t)(\operatorname{tr} A)(t) + (\Delta g(t))^2 (\det A)(t)$$
$$= 1 + \Delta g(t)\left((\operatorname{tr} A)(t) + \Delta g(t)(\det A)(t)\right), \quad t \in I,$$

i.e.,

$$\det(\mathscr{I} + \Delta g(t)A(t))$$
$$= 1 + \Delta g(t)\left((\operatorname{tr} A)(t) + \Delta g(t)(\det A)(t)\right), \quad t \in I. \quad (7.1)$$

Suppose A is Stieltjes regressive. Then

$$\det(\mathscr{I} + \Delta g(t)A(t)) \neq 0, \quad t \in I.$$

In view of (7.1), we obtain

$$1 + \Delta g(t)\left((\operatorname{tr} A)(t) + \Delta g(t)(\det A)(t)\right) \neq 0, \quad t \in I, \quad (7.2)$$

i.e.,

$$\operatorname{tr} A + \Delta g(t) \det A, \quad t \in I,$$

is Stieltjes regressive. Conversely, suppose

$$\operatorname{tr} A + \Delta g(t) \det A, \quad t \in I,$$

is Stieltjes regressive. Then, (7.2) holds. From (7.1), we conclude that A is Stieltjes regressive. This completes the proof. $\square$

Definition 7.4. Assume that A and B are Stieltjes regressive matrix-valued functions on I. Then we define $A \oplus_g B$, $\ominus_g A$, and $A \ominus_g B$ by

$$(A \oplus_g B)(t) = A(t) + B(t) + \Delta g(t)A(t)B(t),$$
$$(\ominus_g A)(t) = -(I + \Delta g(t)A(t))^{-1}A(t), \quad t \in I,$$

and
$$(A \ominus_g B)(t) = (A \oplus_g (\ominus_g B))(t), \quad t \in I,$$
respectively.

Example 7.6. Let $I = \mathbb{R}$. Define $g(t) = \frac{2}{3}t + 1$,
$$A(t) = \begin{pmatrix} 1 & t \\ 2 & 3t \end{pmatrix}, \quad \text{and} \quad B(t) = \begin{pmatrix} t & 1 \\ 2t & 3 \end{pmatrix}, \quad t \in I.$$

Then
$$(A \oplus_g B)(t) = A(t) + B(t) + \Delta g(t) A(t) B(t)$$
$$= \begin{pmatrix} 1 & t \\ 2 & 3t \end{pmatrix} + \begin{pmatrix} t & 1 \\ 2t & 3 \end{pmatrix}$$
$$= \begin{pmatrix} 1+t & 1+t \\ 2(1+t) & 3(1+t) \end{pmatrix}, \quad t \in I$$

and
$$(B \oplus_g A)(t) = A(t) + B(t) + \Delta g(t) B(t) A(t)$$
$$= \begin{pmatrix} 1 & t \\ 2 & 3t \end{pmatrix} + \begin{pmatrix} t & 1 \\ 2t & 3 \end{pmatrix}$$
$$= \begin{pmatrix} 1+t & 1+t \\ 2(1+t) & 3(1+t) \end{pmatrix}, \quad t \in I.$$

Also,
$$\mathscr{I} + \Delta g(t) B(t) = \begin{pmatrix} 1 & 0 \\ 0 & 1 \end{pmatrix}, \quad t \in I$$

and
$$\det(\mathscr{I} + \Delta g(t) B(t)) = 1, \quad t \in I.$$

Hence
$$(\mathscr{I} + \Delta g(t) B(t))^{-1} = \begin{pmatrix} 1 & 0 \\ 0 & 1 \end{pmatrix}, \quad t \in I,$$

and
$$(\ominus_g B)(t) = -(\mathscr{I} + \Delta g(t) B(t))^{-1} B(t)$$
$$= \begin{pmatrix} -t & -1 \\ -2t & -3 \end{pmatrix}, \quad t \in I.$$

Exercise 7.4. Let $I = \mathbb{R}$. Define $g(t) = 2t$,
$$A(t) = \begin{pmatrix} 1 & 1 \\ 2 & -1 \end{pmatrix}, \quad \text{and} \quad B(t) = \begin{pmatrix} 3 & 4 \\ 1 & 0 \end{pmatrix}, \quad t \in I.$$

Then find

(1) $(\ominus_g A)(t), t \in I$,
(2) $(A \oplus_g B)(t), t \in I$.

Answer 7.2.

(1)
$$(\ominus_g A)(t) = \begin{pmatrix} -1 & -1 \\ -2 & 1 \end{pmatrix},$$

(2)
$$(A \oplus_g B)(t) = \begin{pmatrix} 4 & 5 \\ 3 & -1 \end{pmatrix}, \quad t \in I.$$

Theorem 7.7. *The set $\mathscr{R}_g$ under the operation $\oplus_g$ is a group.*

Proof. Let $A, B, C \in \mathscr{R}_g$. Then
$$(\mathscr{I} + \Delta g(t) A)^{-1}, \quad (\mathscr{I} + \Delta g(t) B)^{-1}, \quad (\mathscr{I} + \Delta g(t) C)^{-1}, \quad t \in I,$$
exist, and
$$\mathscr{I} + \Delta g(t)(A \oplus_g B) = \mathscr{I} + \Delta g(t)(A + B + \Delta g(t) AB)$$
$$= \mathscr{I} + \Delta g(t) A + \Delta g(t) B + \Delta g(t)^2 AB$$
$$= \mathscr{I} + \Delta g(t) A + (\mathscr{I} + \Delta g(t) A) \Delta g(t) B$$
$$= (\mathscr{I} + \Delta g(t) A)(\mathscr{I} + \Delta g(t) B), \quad t \in I.$$

Therefore
$$(\mathscr{I} + \Delta g(t)(A \oplus_g B))^{-1}, \quad t \in I,$$
exists, and so, $A \oplus_g B \in \mathscr{R}_g$. Also,

$(A \oplus_g B) \oplus_g C$
$= (A \oplus_g B) + C + \Delta g(t)(A \oplus_g B)C$
$= A + B + \Delta g(t)AB + C + \Delta g(t)(A + B + \Delta g(t)AB)C$
$= A + B + \Delta g(t)AB + C + \Delta g(t)AC + \Delta g(t)BC + \Delta g(t)^2 ABC,$
$\quad t \in I,$

and

$A \oplus_g (B \oplus_g C)$
$= A + (B \oplus_g C) + \Delta g(t)A(B \oplus_g C)$
$= A + B + C + \Delta g(t)BC + \Delta g(t)A(B + C + \Delta g(t)BC)$
$= A + B + C + \Delta g(t)BC + \Delta g(t)AB + \Delta g(t)AC + \Delta g(t)^2 ABC,$
$\quad t \in I.$

Consequently,
$$(A \oplus_g B) \oplus_g C = A \oplus_g (B \oplus_g C),$$
i.e., in $\mathscr{R}_g$, the associative law holds with respect to $\oplus_g$. Note that
$$O \oplus_g A = A \oplus_g O$$
$$= A,$$
i.e., O is an identity element in $\mathscr{R}_g$ with respect to $\oplus_g$. Next,

$A \oplus_g (-(\mathscr{I} + \Delta g(t)A)^{-1}A) = A - (\mathscr{I} + \Delta g(t)A)^{-1}A$
$\qquad - \Delta g(t)(\mathscr{I} + \Delta g(t)A)^{-1}A^2$
$= A - (\mathscr{I} + \Delta g(t)A)^{-1}(\mathscr{I} + \Delta g(t)A)A$
$= A - A$
$= O, \quad t \in I,$

i.e., $-(\mathscr{I} + \Delta g(t)A)^{-1}A$, $t \in I$, is the additive inverse of A with respect to $\oplus_g$. Note that

$$\begin{aligned}\mathscr{I} + \Delta g(t)(-(\mathscr{I} + \Delta g(t)A)^{-1}A) &= (\mathscr{I} + \Delta g(t)A)^{-1}(\mathscr{I} + \Delta g(t)A) \\ &\quad - (\mathscr{I} + \Delta g(t)A)^{-1}\Delta g(t)A \\ &= (\mathscr{I} + \Delta g(t)A)^{-1}, \quad t \in I,\end{aligned}$$

and then $-(\mathscr{I} + \Delta g(t)A)^{-1}A \in \mathscr{R}_g$, $t \in I$. Thus, $(\mathscr{R}_g, \oplus_g)$ is a group. This completes the proof. □

Henceforth, the conjugate of matrix A will be denoted by $\overline{A}$ and the transpose of matrix A will be denoted by A^T. The conjugate transpose of matrix of A will be denoted by $A^* = (\overline{A})^T$.

Theorem 7.8. *Let A and B be Stieltjes regressive matrix-valued functions. Then we have the following:*

(1) A^* *is regressive;*

(2) $\ominus_g A^* = (\ominus_g A)^*;$

(3) $(A \oplus_g B)^* = B^* \oplus_g A^*.$

Proof. Since A is Stieltjes regressive, $(\mathscr{I} + \Delta g(t)A)^{-1}$ exists.

(1) We have

$$\begin{aligned}\mathscr{I} &= (\mathscr{I} + \Delta g(t)A)(\mathscr{I} + \Delta g(t)A)^{-1} \\ &= (\mathscr{I} + \Delta g(t)\overline{A})\overline{(\mathscr{I} + \Delta g(t)A)^{-1}} \\ &= \left((\mathscr{I} + \Delta g(t)A)^{-1}\right)^*(\mathscr{I} + \Delta g(t)A^*), \quad t \in I.\end{aligned}$$

Therefore, both

$$(\mathscr{I} + \Delta g(t)A^*)^{-1} \quad \text{and} \quad (\mathscr{I} + \Delta g(t)A^T)^{-1}, \quad t \in I,$$

exist, and

$$\begin{aligned}(\mathscr{I} + \Delta g(t)A^*)^{-1} &= \left((\mathscr{I} + \Delta g(t)A)^{-1}\right)^*, \\ (\mathscr{I} + \Delta g(t)A^T)^{-1} &= \left((\mathscr{I} + \Delta g(t)A)^{-1}\right)^T, \quad t \in I.\end{aligned}$$

Consequently, A^* is Stieltjes regressive.

(2) We have
$$\begin{aligned}(\ominus_g A)^* &= -\big((\mathscr{I}+\Delta g(t)A)^{-1}A\big)^* \\ &= -A^*\big((\mathscr{I}+\Delta g(t)A)^{-1}\big)^* \\ &= -A^*(\mathscr{I}+\Delta g(t)A^*)^{-1} \\ &= \ominus_g A^*, \quad t \in I.\end{aligned}$$

Thus,
$$(\ominus_g A)^* = \ominus_g A^*.$$

(3) We have
$$\begin{aligned}(A \oplus B)^* &= (A + B + \Delta g(t)AB)^* \\ &= A^* + B^* + \Delta g(t)B^*A^* \\ &= B^* + A^* + \Delta g(t)B^*A^* \\ &= B^* \oplus_g A^*, \quad t \in I.\end{aligned}$$

Thus,
$$(A \oplus B)^* = B^* \oplus_g A^*.$$

This completes the proof. □

7.2 The Stieltjes Matrix Exponential Function

Suppose that A is a continuous $n \times n$-matrix on I and $t_0, T \in I$ with $t_0 < T$. Set
$$\widetilde{A}(t) = \begin{cases} A(t) & \text{for } t \in [t_0, T] \setminus D_g, \\ \dfrac{\log(1 + A(t)\Delta g(t))}{\Delta g(t)} & \text{for } t \in [t_0, T] \cap D_g. \end{cases}$$

Definition 7.5. Define the Stieltjes matrix exponential function as follows:
$$e_{A,g}(t, t_0) = \exp\left[\int_{t_0}^{t} \widetilde{A}(s)d_g s\right], \quad t \in [t_0, T].$$

Example 7.7. Let $I = [0, \infty)$. Define $g(t) = 1 + t$ and
$$A(t) = \begin{pmatrix} -2t & 0 \\ 0 & 3t^2 \end{pmatrix}, \quad t \in I.$$

Then
$$\widetilde{A}(t) = A(t), \quad t \in I,$$

and
$$\begin{aligned} e_{A,g}(t,0) &= \exp\left[\int_{t_0}^{t} \widetilde{A}(s) d_g s\right] \\ &= \exp\left[\int_0^t \widetilde{A}(s) g'(s) ds\right] \\ &= \exp\left[\int_0^t \begin{pmatrix} -2s & 0 \\ 0 & 3s^2 \end{pmatrix} ds\right] \\ &= \exp\left[\begin{pmatrix} -2\int_0^t s\, ds & 0 \\ 0 & 3\int_0^t s^2 ds \end{pmatrix}\right] \\ &= \exp\left[\begin{pmatrix} -t^2 & 0 \\ 0 & t^2 \end{pmatrix}\right], \quad t \in I. \end{aligned}$$

Thus,
$$e_{A,g}(t,0) = \begin{pmatrix} e^{-t^2} & 1 \\ 1 & e^{t^3} \end{pmatrix}, \quad t \in I.$$

In the following, we deduct some of the properties of the Stieltjes-matrix exponential function.

Theorem 7.9. *Suppose A and B are continuous $n \times n$-matrices on I and $t_0 \in I$. Then we have the following:*

(1) $e_{A,g}(t_0, t_0) = 1$, $t \in I$;
(2) $(e_{A,g})'_g(t, t_0) = A(t) e_{A,g}(t, t_0)$, $t \in I$;

(3) $\frac{1}{e_{A,g}(t,t_0)} = e_{\ominus_g A, g}(t, t_0)$, $t \in I$;

(4) $e_{A,g}(t, t_0) e_{B,g}(t, t_0) = e_{A \oplus_g B, g}(t, t_0)$, $t \in I$, provided that A and B commute;

(5) $\frac{e_{A,g}(t,t_0)}{e_{B,g}(t,t_0)} = e_{A \ominus_g B, g}(t, t_0)$, $t \in I$, provided that A and B commute and $\det B \neq 0$ on I.

Proof. (1) We have

$$e_{A,g}(t_0, t_0) = \exp\left[\int_{t_0}^{t_0} \widetilde{A}(s) d_g s\right]$$
$$= 1.$$

Thus,

$$e_{A,g}(t_0, t_0) = 1.$$

(2) We have

$(e_{A,g})'_g(t, t_0)$

$= \left(\exp\left[\int_{t_0}^{t} \widetilde{A}(s) d_g s\right]\right)'_g$

$= e_{A,g}(t, t_0) \left(\int_{[t_0,t) \setminus D_g} A(s) d_g s \right.$

$\left. + \int_{[t_0,t) \cap D_g} \frac{\log(1 + A(s) \Delta g(s))}{\Delta g(s)} d_g s \right)'_g$

$= e_{A,g}(t, t_0) \left(\int_{[t_0,t) \setminus D_g} A(s) d_g s \right.$

$\left. + \sum_{t_k < t} \frac{\log(1 + A(t_k) \Delta g(t_k))}{\Delta g(t_k)} \Delta g(t_k) \right)'_g$

$= e_{A,g}(t, t_0) \left(\int_{[t_0,t) \setminus D_g} A(s) d_g s + \sum_{t_k < t} \log(1 + A(t_k) \Delta g(t_k))\right)'_g$,

$t \in I$,

where t_k denotes the points of discontinuity of g on I. Thus,

$$(e_{A,g})'_g(t,t_0) = A(t)e_{A,g}(t,t_0), \quad t \in I.$$

(3) We have

$$\frac{1}{e_{A,g}(t,t_0)}$$

$$= \frac{1}{\exp\left[\int_{t_0}^t \tilde{A}(s)d_g s\right]}$$

$$= \exp\left[-\int_{t_0}^t \tilde{A}(s)d_g s\right]$$

$$= \exp\left[-\int_{[t_0,t)\setminus D_g} A(s)d_g s - \int_{[t_0,t)\cap D_g} \tilde{A}(s)d_g s\right]$$

$$= \exp\left[\int_{[t_0,t)\setminus D_g} \ominus_g A(s)d_g s - \sum_{t_k<t} \frac{\log(1+A(t_k)\Delta g(t_k))}{\Delta g(t_k)}\Delta g(t_k)\right]$$

$$= \exp\left[\int_{[t_0,t)\setminus D_g} \ominus_g A(s)d_g s - \sum_{t_k<t} \frac{\log\left(\frac{1}{1+A(t_k)\Delta g(t_k)}\right)}{\Delta g(t_k)}\Delta g(t_k)\right]$$

$$= \exp\left[\int_{[t_0,t)\setminus D_g} \ominus_g A(s)d_g s - \sum_{t_k<t} \frac{\log(1+(\ominus_g A(t_k))\Delta g(t_k))}{\Delta g(t_k)}\Delta g(t_k)\right]$$

$$= \exp\left[\int_{[t_0,t)\setminus D_g} \ominus_g A(s)d_g s + \int_{[t_0,t)\cap D_g} \log(1+(\ominus_g A)(s)\Delta g(s))d_g s\right]$$

$$= \exp\left[\int_{t_0}^t \ominus_g \tilde{A}(s)d_g s\right], \quad t \in I,$$

where t_k denotes the points of discontinuity of g on I. Thus,

$$\frac{1}{e_{A,g}(t,t_0)} = e_{\ominus_g A,g}(t,t_0), \quad t \in I.$$

(4) We have

$$e_{A,g}(t,t_0)e_{B,g}(t,t_0), \quad t \in I,$$

$e_{A,g}(t,t_0)e_{B,g}(t,t_0)$

$$= \left(\exp\left[\int_{[t_0,t)\setminus D_g} A(s)d_g s + \int_{[t_0,t)\cap D_g} \frac{\log(1+A(s)\Delta g(s))}{\Delta g(s)} d_g s\right]\right)$$

$$\times \left(\exp\left[\int_{[t_0,t)\setminus D_g} B(s)d_g s + \int_{[t_0,t)\cap D_g} \frac{\log(1+B(s)\Delta g(s))}{\Delta g(s)} d_g s\right]\right)$$

$$= \exp\left[\int_{[t_0,t)\setminus D_g} (A(s)+B(s))d_g s + \int_{[t_0,t)\cap D_g} \frac{\log(1+A(s)\Delta g(s))+\log(1+B(s)\Delta g(s))}{\Delta g(s)} d_g s\right]$$

$$= \exp\left[\int_{[t_0,t)\setminus D_g} (A\oplus_g B)(s)d_g s + \int_{[t_0,t)\cap D_g} \frac{\log(1+(A\oplus_g B)(s)\Delta g(s))}{\Delta g(s)} d_g s\right]$$

$$= e_{(A\oplus_g B),g}(t,t_0), \quad t\in I.$$

Thus,

$$e_{A,g}(t,t_0)e_{B,g}(t,t_0) = e_{(A\oplus_g B),g}(t,t_0), \quad t\in I.$$

(5) Using the identities in points (3) and (4), we find

$$\frac{e_{A,g}(t,t_0)}{e_{B,g}(t,t_0)} = e_{A,g}(t,t_0)e_{\ominus_g B,g}(t,t_0)$$

$$= e_{A\oplus_g(\ominus_g B),g}(t,t_0)$$

$$= e_{A\ominus_g B,g}(t,t_0), \quad t\in I.$$

Thus,
$$\frac{e_{A,g}(t,t_0)}{e_{B,g}(t,t_0)} = e_{A\ominus_g B,g}(t,t_0), \quad t \in I.$$

This completes the proof. □

Theorem 7.10 (The Stieltjes–Liouville Formula). *Let $A \in \mathscr{R}_g$ be a 2×2-matrix-valued function and assume that X is a solution of*
$$X'_g = A(t)X, \quad t \in I.$$
Then X satisfies the Stieltjes–Liouville formula
$$\det X(t) = e_{\operatorname{tr} A, g}(t, t_0) \det X(t_0), \quad t \in I.$$

Proof. Since $A \in \mathscr{R}_g$, it follows that $\operatorname{tr} A \mathscr{R}_g$. Let
$$A(t) = \begin{pmatrix} a_{11}(t) & a_{12}(t) \\ a_{21}(t) & a_{22}(t) \end{pmatrix} \quad \text{and} \quad X(t) = \begin{pmatrix} x_{11}(t) & x_{12}(t) \\ x_{21}(t) & x_{22}(t) \end{pmatrix}, \quad t \in I.$$
Then
$$\begin{pmatrix} x'_{11g}(t) & x'_{12g}(t) \\ x'_{21g}(t) & x'_{22g}(t) \end{pmatrix}$$
$$= \begin{pmatrix} a_{11}(t) & a_{12}(t) \\ a_{21}(t) & a_{22}(t) \end{pmatrix} \begin{pmatrix} x_{11}(t) & x_{12}(t) \\ x_{21}(t) & x_{22}(t) \end{pmatrix}$$
$$= \begin{pmatrix} a_{11}(t)x_{11}(t) + a_{12}(t)x_{12}(t) & a_{11}(t)x_{12}(t) + a_{12}(t)x_{22}(t) \\ a_{21}(t)x_{11}(t) + a_{22}(t)x_{21}(t) & a_{21}(t)x_{12}(t) + a_{22}(t)x_{22}(t) \end{pmatrix}, \quad t \in I,$$
whereupon
$$\begin{cases} x'_{11g}(t) = a_{11}(t)x_{11}(t) + a_{12}(t)x_{21}(t), \\ x'_{12g}(t) = a_{11}(t)x_{12}(t) + a_{12}(t)x_{22}(t), \\ x'_{21g}(t) = a_{21}(t)x_{11}(t) + a_{22}(t)x_{21}(t), \\ x'_{22g}(t) = a_{21}(t)x_{12}(t) + a_{22}(t)x_{22}(t), \quad t \in I. \end{cases}$$

Then
$$\det X(t) = x_{11}(t)x_{22}(t) - x_{12}(t)x_{21}(t), \quad t \in I,$$

and

$(\det X)'_g(t)$
$= x'_{11g}(t)x_{22}(t) + x_{11}(t)x'_{22g}(t) - x'_{12g}(t)x_{21}(t) - x_{12}(t)x'_{21g}(t)$
$= \det \begin{pmatrix} x'_{11g}(t) & x'_{12g}(t) \\ x_{21}(t) & x_{22}(t) \end{pmatrix} + \det \begin{pmatrix} x_{11}(t) & x_{12}(t) \\ x'_{21g}(t) & x'_{22g}(t) \end{pmatrix}$
$= \det \begin{pmatrix} a_{11}(t)x_{11}(t) + a_{12}(t)x_{21}(t) & a_{11}(t)x_{12}(t) + a_{12}(t)x_{22}(t) \\ x_{21}(t) & x_{22}(t) \end{pmatrix}$
$\quad + \det \begin{pmatrix} x_{11}(t) & x_{12}(t) \\ x'_{21g}(t) & x'_{22g}(t) \end{pmatrix}$
$= [a_{11}(t)x_{11}(t)x_{22}(t) + a_{12}(t)x_{21}(t)x_{22}(t)$
$\quad - a_{11}(t)x_{12}(t)x_{21}(t) - a_{12}(t)x_{21}(t)x_{22}(t)]$
$\quad + \det \begin{pmatrix} x_{11}(t) & x_{12}(t) \\ x'_{21g}(t) & x_{22}(t) \end{pmatrix}$
$= a_{11}(t) \det X(t) + \det \begin{pmatrix} x_{11}(t) & x_{12}(t) \\ x'_{21g}(t) & x'_{22g}(t) \end{pmatrix}$
$= a_{11}(t) \det X(t)$
$\quad + \det \begin{pmatrix} x_{11}(t) & x_{12}(t) \\ a_{21}(t)x_{11}(t) + a_{22}(t)x_{21}(t) & a_{21}(t)x_{12}(t) + a_{22}(t)x_{22}(t) \end{pmatrix}$
$= a_{11}(t) \det X(t) + [a_{21}(t)x_{11}(t)x_{12}(t) + a_{22}(t)x_{11}(t)x_{22}(t)$
$\quad - a_{21}(t)x_{11}(t)x_{12}(t) - a_{22}(t)x_{21}(t)x_{12}(t)]$
$= a_{11}(t) \det X(t) + a_{22}(t) \det X(t)$
$= \operatorname{tr} A(t) \det X(t), \quad t \in I,$

i.e.,
$$(\det X)'_g(t) = \operatorname{tr} A(t) \det X(t), \quad t \in I.$$

Thus,
$$\det X(t) = e_{\operatorname{tr} A,g}(t,t_0) \det X(t_0), \quad t \in I.$$

This completes the proof. □

7.3 nth-Order Stieltjes Differential Equations

Theorem 7.11. *A function x solves the nth-order Stieltjes differential equation*
$$x_g^{(n)} + a_1(t)x_g^{(n-1)} + \cdots + a_{n-1}(t)x'_g + a_n(t)x = f(t), \quad t \in I, \quad (7.3)$$
$t \in I$, $a_i \in \mathscr{C}(I)$, $i \in \{1,\ldots,n\}$, if and only if the functions
$$\begin{aligned} y_1 &= x, \\ y_2 &= x'_g, \\ &\vdots \\ y_n &= x_g^{(n-1)} \end{aligned} \qquad (7.4)$$

satisfy the system
$$\begin{aligned} y'_{1g} &= y_2, \\ y'_{2g} &= y_3, \\ &\vdots \\ y'_{ng} &= -a_1(t)y_{n-1} - \cdots - a_{n-1}(t)y_2 - a_n(t)y_1 + f(t), \quad t \in I, \end{aligned} \qquad (7.5)$$

Proof. Suppose x is a solution to (7.3). We introduce the functions (7.4). Then,
$$\begin{aligned} y'_{1g} &= y_2, \\ y'_{2g} &= y_3, \end{aligned}$$

$$\vdots$$
$$y'_{n-1\,g} = y_n.$$

Hence, from (7.3), we get (7.5).

Conversely, suppose $y_1, \ldots, y_n$ satisfy the system (7.5). Differentiating, in the sense of Stieltjes, the first equation of (7.5), we get

$$y''_{1g} = y'_{2g}.$$

Hence, from the second equation of (7.5), we obtain

$$y''_{1g} = y_3,$$

and so on,

$$y_{1g}^{(n-1)} = y_n.$$

From here and from the last equation, denoting $x = y_1$, we obtain (7.3). This completes the proof. □

Example 7.8. Consider the Stieltjes differential equation

$$x'''_g - t^2 x''_g - tx'_g + 2tx = \sin t, \quad t \in \mathbb{R}.$$

Introduce the new unknowns:

$$y_1 = x,$$
$$y_2 = x'_g,$$
$$y_3 = x''_g.$$

Then

$$y'_{1g} = y_2,$$
$$y'_{2g} = y_3,$$
$$y'_{3g} = t^2 y_3 + t y_2 - 2t y_1 + \sin t.$$

Example 7.9. Consider the system

$$x'_{1g} = 2t x_1 - (t^2 + e) x_2 + \sin t,$$
$$x'_{2g} = x_1 + x_2 + \cos t, \quad t \in \mathbb{R}.$$

From its second equation, we get

$$x_1 = x'_{2g} - x_2 - \cos t, \quad t \in \mathbb{R},$$

which we substitute in the first equation, and we obtain

$$x''_{2g} - (e + 2t)x'_{2g} + (t + e)^2 x_2 + 2t \cos t = 1,$$

$t \in \mathbb{R}$.

Exercise 7.5. Prove that a 2×2-constant coefficients linear Stieltjes differential system

$$x'_g = Ax, \quad \text{where } x = \begin{pmatrix} x_1 \\ x_2 \end{pmatrix},$$

can be written as the second-order Stieltjes differential equation for y_1 given by

$$x''_{1g} - \operatorname{tr}(A)x'_{1g} + \det(A)x_1 = 1.$$

Definition 7.6. For a system of functions $x_1, \ldots, x_n \in \mathscr{C}_g^n(I)$, define the Stieltjes–Wronskian as follows:

$$W_g(x_1, \ldots, x_n) = \det \begin{pmatrix} x_1 & x_2 & \cdots & x_n \\ x'_{1g} & x'_{2g} & \cdots & x'_{ng} \\ \vdots & \vdots & \vdots & \vdots \\ x_{1g}^{(n-1)} & x_{2g}^{(n-1)} & \cdots & x_{ng}^{(n-1)} \end{pmatrix}.$$

Example 7.10. Let $I = [0, \infty)$. Define

$$g(t) = t + t^2,$$
$$x_1(t) = 1 + t^3,$$

and

$$x_2(t) = t + t^4, \quad t \in I.$$

Then
$$x'_{1g}(t) = \frac{x'_1(t)}{g'(t)}$$
$$= \frac{3t^2}{1+2t}, \quad t \in I,$$

and
$$x'_{2g}(t) = \frac{x'_2(t)}{g'(t)}$$
$$= \frac{1+4t^3}{1+2t}, \quad t \in I.$$

Hence,
$$W_g(x_1, x_2)(t) = \det \begin{pmatrix} x_1(t) & x_2(t) \\ x'_{1g}(t) & x'_{2g}(t) \end{pmatrix}$$
$$= \det \begin{pmatrix} 1+t^3 & t+t^4 \\ \dfrac{3t^2}{1+2t} & \dfrac{1+4t^3}{1+2t} \end{pmatrix}$$
$$= \frac{(1+t^3)(1+4t^3) - 3t^2(t+t^4)}{1+2t}$$
$$= \frac{1 + 4t^3 + t^3 + 4t^6 - 3t^3 - 3t^6}{1+2t}$$
$$= \frac{1 + 2t^3 + t^6}{1+2t}, \quad t \in I.$$

Thus,
$$W_g(x_1, x_2)(t) = \frac{(1+t^3)^2}{1+2t}, \quad t \in I.$$

Theorem 7.12. Let $x_1, \ldots, x_n \in \mathscr{C}_g^n(I)$ be such that
$$W_g(x_1, \ldots, x_n, x)(t) \neq 0, \quad t \in I. \tag{7.6}$$

Then
$$\frac{W_g(x_1, \ldots, x_n, x)(t)}{W_g(x_1, \ldots, x_n)(t)} = 0, \quad t \in I, \tag{7.7}$$

is an nth-order Stieltjes differential equation whose solutions are $x_1, \ldots, x_n$.

Proof. We have

$$W_g(x_1, \ldots, x_n, x)(t) = \det \begin{pmatrix} x_1(t) & \cdots & x_n(t) & x(t) \\ x'_{1g}(t) & \cdots & x'_{ng}(t) & x'_g(t) \\ \vdots & \vdots & \vdots & \vdots \\ x^{(n)}_{1g}(t) & \cdots & x^{(n)}_{ng}(t) & x^{(n)}_g(t) \end{pmatrix}$$

$$= x^{(n)}_g(t) \det \begin{pmatrix} x_1(t) & \cdots & x_n(t) \\ \vdots & \vdots & \vdots \\ x^{(n-1)}_{1g}(t) & \cdots & x^{(n-1)}_{ng}(t) \end{pmatrix}$$

$$+ x^{(n-1)}_g(t) \det \begin{pmatrix} x_1(t) & \cdots & x_n(t) \\ \vdots & \vdots & \vdots \\ x^{(n-2)}_{1g}(t) & \cdots & x^{(n-2)}_{ng}(t) \\ x^{(n)}_{1g}(t) & \cdots & x^{(n)}_{ng}(t) \end{pmatrix}$$

$$+ \cdots$$

$$+ y(t) \det \begin{pmatrix} x'_{1g}(t) & \cdots & x'_{ng}(t) \\ \vdots & \vdots & \vdots \\ x^{(n)}_{1g}(t) & \cdots & x^{(n)}_{ng}(t) \end{pmatrix}$$

$$= x^{(n)}_g(t) W_g(x_1, \ldots, x_n)(t)$$

$$+ x^{(n-1)}_g(t) \det \begin{pmatrix} x_1(t) & \cdots & x_n(t) \\ \vdots & \vdots & \vdots \\ x^{(n-2)}_{1g}(t) & \cdots & x^{(n-2)}_{ng}(t) \\ x^{(n)}_{1g}(t) & \cdots & x^{(n)}_{ng}(t) \end{pmatrix}$$

$$+ \cdots$$

$$+ x(t) \det \begin{pmatrix} x'_{1g}(t) & \cdots & x'_{ng}(t) \\ \vdots & \vdots & \vdots \\ x^{(n)}_{1g}(t) & \cdots & x^{(n)}_{ng}(t) \end{pmatrix}, \quad t \in I.$$

Hence, in view of (7.7), we obtain

$$\frac{W_g(x_1, \ldots, x_n, x)(t)}{W_g(x_1, \ldots, x_n)(t)}$$

$$= x^{(n)}_g(t) + \frac{x^{(n-1)}_g(t)}{W_g(x_1, \ldots, x_n)(t)} \det \begin{pmatrix} x_1(t) & \cdots & x_n(t) \\ \vdots & \vdots & \vdots \\ x^{(n-2)}_{1g}(t) & \cdots & x^{(n-2)}_{ng}(t) \\ x^{(n)}_{1g}(t) & \cdots & x^{(n)}_{ng}(t) \end{pmatrix}$$

$$+ \cdots +$$

$$+ \frac{x(t)}{W_g(x_1, \ldots, x_n)(t)} \det \begin{pmatrix} x'_{1g}(t) & \cdots & x'_{ng}(t) \\ \vdots & \vdots & \vdots \\ x^{(n)}_{1g}(t) & \cdots & x^{(n)}_{ng}(t) \end{pmatrix} = 0, \quad t \in I,$$

which is an nth-order Stieltjes differential equation because (7.6) holds. Next,

$$W_g(x_1, \ldots, x_n, x_j)(t) = \begin{pmatrix} x_1(t) & \cdots & x_j(t) & \cdots & x_n(t) & x_j(t) \\ x'_{1g}(t) & \cdots & x'_{jg}(t) & \cdots & x'_{ng}(t) & x'_{jg}(t) \\ \vdots & \vdots & \vdots & \vdots & \vdots & \vdots \\ x^{(n)}_{1g}(t) & \cdots & x^{(n)}_{jg}(t) & \cdots & x^{(n)}_{ng}(t) & x^{(n)}_{jg}(t) \end{pmatrix}$$

$$= 0, \quad t \in I, \quad j \in \{1, \ldots, n\}.$$

Therefore
$$\frac{W_g(x_1,\ldots,x_n,x_j)(t)}{W_g(x_1,\ldots,x_n)(t)} = 0, \quad t \in I, \quad j \in \{1,\ldots,n\}.$$

Thus, x_j, $j \in \{1,\ldots,n\}$, is a solution to (7.7). This completes the proof. $\square$

7.4 Stieltjes Homogeneous Systems

Theorem 7.13 (Superposition Principle). *If the vector-valued functions x_1 and x_2 are solutions to the homogeneous Stieltjes differential system*

$$x'_g = A(t)x, \quad t \in I, \tag{7.8}$$

then any linear combination

$$x = ax_1 + bx_2, \quad a, b \in \mathbb{R},$$

is also a solution to (7.8).

Proof. We have
$$(x_1)'_g = A(t)x_1,$$
$$(x_2)'_g = A(t)x_2, \quad t \in I.$$

Then
$$\begin{aligned}
x'_g &= (ax_1 + bx_2)'_g \\
&= a(x_1)'_g + b(x_2)'_g \\
&= aA(t)x_1 + bA(t)x_2 \\
&= A(t)(ax_1) + A(t)(bx_2) \\
&= A(t)(ax_1 + bx_2) \\
&= A(t)x, \quad t \in I.
\end{aligned}$$

Thus,
$$x'_g = A(t)x, \quad t \in I.$$

This completes the proof. $\square$

Definition 7.7. The set of functions $\{x_1, \ldots, x_n\}$ is a fundamental set of solutions for the system (7.8) on I if and only if the following holds:

(1) $(x_i)'_g = A(t)x_i$, $1 \leq i \leq n$, for any $t \in I$,

(2) the set $\{x_1, \ldots, x_n\}$ is linearly independent in I.

Definition 7.8. The general solution of the system (7.8) on I is a vector-valued function x_{gen} that can be written as a linear combination

$$x_{\text{gen}} = c_1 x_1(t) + \cdots + c_n x_n(t), \quad t \in I,$$

where $\{x_1, \ldots, x_n\}$ is a fundamental set of solutions of (7.8) on I, while $c_1, \ldots, c_n$ are arbitrary constants.

Definition 7.9. Let $\{x_1, \ldots, x_n\}$ be a fundamental set of solutions of (7.8) on I. Then the matrix

$$X(t) = [x_1(t), \ldots, x_n(t)], \quad t \in I,$$

is called a fundamental matrix.

The Stieltjes–Wronskian of the set $\{x_1, \ldots, x_n\}$ is the function

$$W_g(t) = \det(X(t)), \quad t \in I.$$

Theorem 7.14. *Let $x_1, \ldots, x_n$ be linearly independent solutions to the system (7.8) in I. Then*

$$W_g(t) \neq 0, \quad t \in I. \tag{7.9}$$

Proof. Assume that

$$W_g(t_0) = 0$$

for some $t_0 \in I$. Then, there are constants $c_1, \ldots, c_n$ such that

$$c_1 x_1(t_0) + \cdots + c_n x_n(t_0) = 0$$

and

$$(c_1, \ldots, c_n) \neq (0, \ldots, 0). \tag{7.10}$$

Since
$$x(t) = c_1 x_1(t) + \cdots + c_n x_n(t), \quad t \in I,$$
is a solution to the system (7.8) for which
$$x(t_0) = 0,$$
we get that
$$x(t) = 0, \quad t \in I,$$
and
$$c_1 x_1(t) + \cdots + c_n x_n(t) = 0, \quad t \in I,$$
which is impossible because $x_1, \ldots, x_n$ are linearly independent in I and (7.10) holds. Thus, (7.9) holds. This completes the proof. □

Remark 7.1. By Theorem 7.14, it follows that the solutions $x_1, \ldots, x_n$ of the system (7.8) are linearly independent if and only if
$$W_g(t_0) \neq 0$$
for some $t_0 \in I$, i.e., the vectors
$$x_1(t_0), \ldots, x_n(t_0)$$
are linearly independent. Let
$$e_1 = (1, 0, \ldots, 0),$$
$$e_2 = (0, 1, \ldots, 0),$$
$$\vdots$$
$$e_n = (0, 0, \ldots, 1).$$
Note that $e_1, \ldots, e_n$ are linearly independent in I. Hence, the solutions $x_1, \ldots, x_n$ of the system (7.8) for which
$$x_i(t_0) = i, \quad i \in \{1, \ldots, n\},$$

are linearly independent. Therefore, the system (7.8) has exactly n linearly independent solutions.

Example 7.11. Let I, g, x_1 and x_2 be as in Example 7.10. We have that x_1 and x_2 are linearly independent. By the computations in Example 7.10, we have

$$W_g(x_1, x_2)(t) = \frac{(1+t^3)^2}{1+2t}$$
$$\neq 0, \quad t \in I.$$

Theorem 7.15 (The Stieltjes–Abel Theorem II). *Let A be an $n \times n$ matrix-valued continuous function on a domain $I \subset \mathbb{R}$ and $t_0 \in I$ be arbitrarily chosen. Then the Stieltjes–Wronskian $W_g(t)$ of the system (7.8) satisfies*

$$W_g(t) = W_g(t_0) \int_{t_0}^x \operatorname{tr}(A(\tau)) d_g \tau, \quad t \in I. \tag{7.11}$$

Proof. Note that for every $n \times n$ matrix-valued Stieltjes differentiable, invertible function B, we have the identity

$$(\det B)'_g(t) = \det(B(t)) \operatorname{tr}\left(B^{-1}(t) B'_g(t)\right).$$

Hence, we get

$$\begin{aligned} W'_g(t) &= \det(X(t)) \operatorname{tr}\left(X^{-1}(t) X'_g(t)\right) \\ &= W_g(t) \operatorname{tr}\left(X^{-1}(t) A(t) X(t)\right) \\ &= W_g(t) \operatorname{tr}\left(X^{-1}(t) X(t) A(t)\right) \\ &= W_g(t) \operatorname{tr}(A(t)), \end{aligned}$$

i.e.,

$$W'_g(t) = W_g(t) \operatorname{tr}(A(t)), \quad t \in I.$$

Now, integrating the last equation, in the sense of Stieltjes, from t_0 to t, we obtain (7.11). This completes the proof. □

Example 7.12. Let $I = [0, \infty)$, $g(t) = 1 + t^3$, $t \in I$, and

$$A(t) = \begin{pmatrix} -1 & 1 & 2 \\ -3 & t & 5 \\ 0 & t^2 + t & t^2 \end{pmatrix}, \quad t \in I.$$

Then

$$\operatorname{tr}(A(t)) = -1 + t + t^2, \quad t \in I,$$

and

$$\begin{aligned}
\int_0^t \operatorname{tr}(A(\tau)) d_g \tau &= \int_0^t (-1 + \tau + \tau^2) g'(\tau) d\tau \\
&= 3 \int_0^x (-1 + \tau + \tau^2) \tau^2 d\tau \\
&= 3 \left(-\int_0^t \tau^2 d\tau + \int_0^t \tau^3 d\tau + \int_0^t \tau^4 d\tau \right) \\
&= 3 \left(-\frac{t^3}{3} + \frac{t^4}{4} + \frac{t^5}{5} \right), \quad t \in I.
\end{aligned}$$

Hence,

$$W_g(t) = 3 W_g(0) \left(-\frac{t^3}{3} + \frac{t^4}{4} + \frac{t^5}{5} \right), \quad t \in I.$$

Theorem 7.16. *Let A be an $n \times n$-matrix-valued continuous function on a domain $I \subset \mathbb{R}$ and $\{x_1, \ldots, x_n\}$ be a fundamental set of solutions for the system (7.8) on I. Also, let x be a solution to the system (7.8) defined on I. Then there exist unique constants $c_1, \ldots, c_n$ such that*

$$x(t) = c_1 x_1(t) + \cdots + c_n x_n(t), \quad t \in I. \tag{7.12}$$

Proof. We have that the only solution of (7.8) that takes the value $x(t_0)$ at $t_0 \in I$ is x. We will find constants $c_1, \ldots, c_n$ that are solutions of the algebraic linear equation

$$x(t_0) = c_1 x_1(t_0) + \cdots + c_n x_n(t_0).$$

Introducing the notation

$$X(t) = [x_1(t), \ldots, x_n(t)], \quad t \in I,$$

and

$$\begin{pmatrix} c_1 \\ \vdots \\ c_n \end{pmatrix},$$

the above algebraic linear equation has the form

$$x(t_0) = X(t_0)c. \tag{7.13}$$

Since $\{x_1, \ldots, x_n\}$ is a fundamental set of solutions of the system (7.8), we have that

$$\det X(t_0) \neq 0.$$

Therefore, there exist unique constants $c_1, \ldots, c_n$ satisfying the system (7.13). By the superposition principle, Theorem 7.13, it follows that

$$c_1 x_1(t) + \cdots + c_n x_n(t), \quad t \in I,$$

is also a solution of (7.8). Hence, we obtain that x has the form (7.12). This completes the proof. $\square$

7.5 Stieltjes Fundamental Matrix Solutions

Suppose that A is an $n \times n$-matrix whose entries $a_{ij} \in \mathscr{C}(I)$, $i, j \in \{1, \ldots, n\}$. Also, let $t_0 \in I$.

Definition 7.10. The matrix $\Phi(\cdot, t_0)$ is said to be the Stieltjes transitive matrix for the system (7.8) provided it is the continuous solution of the SIVP

$$\begin{aligned} \Phi'_g(t, t_0) &= A(t)\Phi(t, t_0), \quad t \in I, \\ \Phi(t_0, t_0) &= \mathscr{I}, \end{aligned} \tag{7.14}$$

where
$$\mathscr{I} = \begin{pmatrix} 1 & 0 & \cdots & 0 \\ 0 & 1 & \cdots & 0 \\ \vdots & \vdots & \vdots & \vdots \\ 0 & 0 & \cdots & 1 \end{pmatrix}.$$

Instead of $\Phi(t, t_0)$, we usually write $\Phi(t)$ for the Stieltjes transitive matrix.

Theorem 7.17. *Let B be a constant matrix. Then the matrix*
$$\Phi(t, t_0) = e_{B,g}(t, t_0), \quad t \in I,$$
is the Stieltjes transitive matrix for the system
$$x'_g = Bx, \quad t \in I.$$

Proof. We have
$$\Phi(t_0, t_0) = e_{B,g}(t_0, t_0)$$
$$= \mathscr{I}$$
and
$$\Phi'_g(t, t_0) = B e_{B,g}(t, t_0)$$
$$= B\Phi(t, t_0), \quad t \in I.$$
This completes the proof. $\square$

Theorem 7.18. *Let Ψ be a fundamental matrix for the system (7.8). Then, for any constant nonsingular $n \times n$-matrix C, the matrix*
$$\Psi C \qquad (7.15)$$
is also a fundamental matrix for (7.8). Moreover, any fundamental matrix of (7.8) is of the form (7.15) for some constant nonsingular $n \times n$-matrix C.

Proof. Let C be a constant nonsingular $n \times n$-matrix. Since Ψ is a fundamental matrix of (7.8), we have
$$\Psi'_g(t) = A(t)\Psi(t), \quad t \in I.$$

Then
$$(\Psi C)'_g(t) = \Psi'_g(t)C$$
$$= A(t)\Psi(t)C$$
$$= A(t)(\Psi(t)C), \quad t \in I.$$
Thus, Ψ and ΨC are solutions to the system (7.8). Since
$$\det \Psi(t) \neq 0, \quad t \in I,$$
and
$$\det C \neq 0,$$
we get
$$\det(\Psi(t)C) = \det \Psi(t) \det C$$
$$\neq 0, \quad t \in I.$$
Therefore, ΨC is a fundamental matrix for the system (7.8). Now, let Ψ_1 and Ψ_2 be two fundamental matrices for the system (7.8). Then, we have
$$\det \Psi_1(t) \neq 0,$$
$$\det \Psi_2(t) \neq 0, \quad t \in I, \tag{7.16}$$
and
$$\Psi'_{1g}(t) = A(t)\Psi_1(t),$$
$$\Psi'_{2g}(t) = A(t)\Psi_2(t), \quad t \in I.$$
Also, let
$$C_1(t) = \Psi_2^{-1}(t)\Psi_1(t), \quad t \in I.$$
Hence
$$\Psi_1(t) = \Psi_2(t)C_1(t), \quad t \in I, \tag{7.17}$$
and
$$\Psi'_{1g}(t) = \Psi'_{2g}(t)C_1(t) + \Psi_2(t)C'_{1g}(t)$$
$$= A(t)\Psi_2(t)C_1(t) + \Psi_2(t)C'_{1g}(t)$$
$$= A(t)\Psi_1(t) + \Psi_2(t)C'_{1g}(t), \quad t \in I.$$

Therefore,
$$\Psi_2(t)C'_{1g}(t) = \mathscr{I}, \quad t \in I.$$

Using the second relation of (7.16), we get
$$C'_{1g}(t) = \mathscr{I}, \quad t \in I.$$

Thus, C_1 is a constant matrix. From (7.17), we have
$$\det(\Psi_1(t)) = \det(\Psi_2(t)C_1(t))$$
$$= \det(\Psi_2(t))\det(C_1(t)), \quad t \in I.$$

Now, applying (7.16), we find
$$\det(C_1(t)) \neq 0, \quad t \in I.$$

Therefore, C_1 is a constant nonsingular matrix. This completes the proof. □

Corollary 7.1. *Let $\Phi(\cdot, t_0)$ be the transition matrix for the system (7.8) and Ψ be a fundamental matrix for the system (7.8). Then*
$$\Phi(t, t_0) = \Psi(t)\Psi^{-1}(t_0), \quad t \in I. \tag{7.18}$$

Moreover, the solution of the SIVP for (7.8) is given by
$$x(t) = \Psi(t)\Psi^{-1}(t_0)x_0, \quad t \in I, \tag{7.19}$$

where
$$x_0 = \begin{pmatrix} x_{01} \\ \vdots \\ x_{0n} \end{pmatrix}.$$

Proof. Since Ψ is a fundamental matrix for the system (7.8), by Theorem 7.18, it follows that the matrix
$$\Psi\Psi^{-1}(t_0) \tag{7.20}$$
is a fundamental matrix for the system (7.8). Next,
$$\Psi(t_0)\Psi^{-1}(t_0) = \mathscr{I}.$$

Thus, the matrix (7.20) is a solution of the SIVP for (7.8). Since the SIVP for (7.8) has a unique solution, we obtain (7.18). Next, let x be defined by (7.19). Then,

$$x(t_0) = \Psi(t_0)\Psi^{-1}(t_0)x_0$$
$$= x_0.$$

Hence, using the fact that the SIVP for (7.8) has a unique solution, we conclude that (7.19) is the unique solution to the SIVP for (7.8). This completes the proof. □

Remark 7.2. Note that any fundamental matrix Ψ of the system (7.8) determines the matrix A in a unique way. Actually, we have

$$\Psi'_g(t) = -A(t)\Psi(t), \quad t \in I,$$

whereupon

$$A(t) = \Psi'_g(t)\Psi^{-1}(t), \quad t \in I.$$

In the general case, the converse is not true. Since, by Theorem 7.18, for any constant nonsingular $n \times n$-matrix C, we have that

$$\Psi(t)C, \quad t \in I,$$

is a fundamental matrix for (7.8), and

$$(\Psi C)'_g(t)(\Psi(t)C)^{-1} = \Psi'_g(t)CC^{-1}\Psi^{-1}(t)$$
$$= \Psi'_g(t)\Psi^{-1}(t)$$
$$= A(t), \quad t \in I.$$

Now, we will deduct some important properties of the Stieltjes transitive matrix for the system (7.8).

Theorem 7.19. *Let $\Phi(\cdot, t_0)$ be the Stieltjes transitive matrix for the system (7.8). Then we have the following:*
(1) $\Phi(t, t_0) = \Phi(t, t_1)\Phi(t_1, t_0)$, $t, t_1 \in I$;
(2) $\Phi^{-1}(t, t_0) = \Phi(t_0, t)$, $t \in I$;
(3) $\Phi(t, t) = \mathscr{I}$, $t \in I$.

Proof. Let Ψ be a fundamental matrix for the system (7.8).
(1) In view of (7.18), we have
$$\Phi(t, t_0) = \Psi(t)\Psi^{-1}(t_0),$$
$$\Phi(t_1, t_0) = \Psi(t_1)\Psi^{-1}(t_0), \quad t, t_1 \in I.$$
Then
$$\Phi(t, t_0) = \Psi(t)\Psi^{-1}(t_0)$$
$$= \Psi(t)\Psi^{-1}(t_1)\Psi(t_1)\Psi^{-1}(t_0)$$
$$= \Phi(t, t_1)\Phi(t_1, t_0), \quad t, t_1 \in I.$$

(2) In view of (7.18), we get
$$\mathscr{I} = \Phi(t_0, t_0)$$
$$= \Phi(t_0, t)\Phi(t, t_0), \quad t \in I,$$
whereupon
$$\Phi^{-1}(t, t_0) = \Phi(t_0, t), \quad t \in I.$$

(3) For any $t \in I$, applying the relation (7.18), we find
$$\Phi(t, t) = \Psi(t)\Psi^{-1}(t)$$
$$= \mathscr{I}, \quad t \in I.$$

This completes the proof. □

Theorem 7.20. *Let $\Phi(\cdot, t_0)$ be the Stieltjes transitive matrix for the system (7.8). Then we have the following:*
(1)
$$\frac{\partial}{\partial_g t_0}\Phi(t, t_0) = -\Phi(t, t_0)A(t_0), \quad t \in I;$$
(2)
$$\Phi(t, t_0) = \mathscr{I} + \int_{t_0}^{t} A(s)\Phi(s, t_0)d_g s, \quad t \in I;$$

(3)
$$\Phi(t,t_0) = \mathscr{I} + \int_{t_0}^{t} \Phi(t,s)A(s)d_g s, \quad t \in I.$$

Proof. (1) By Theorem 7.19, we have
$$\Phi(t,t_0) = \Phi^{-1}(t_0,t), \quad t \in I.$$

Hence,
$$\frac{\partial}{\partial_g t_0}\Phi(t,t_0) = \frac{\partial}{\partial_g t_0}\Phi^{-1}(t_0,t)$$
$$= -\Phi^{-1}(t_0,t)\Phi^{-1}(t_0,t)\Phi'_g(t_0,t)$$
$$= -\Phi^{-1}(t_0,t)\Phi'_g(t_0,t)\Phi^{-1}(t_0,t)$$
$$= -\Phi^{-1}(t_0,t)A(t_0)\Phi(t_0,t)\Phi^{-1}(t_0,t)$$
$$= -\Phi^{-1}(t_0,t)A(t_0), \quad t \in I.$$

Thus,
$$\frac{\partial}{\partial_g t_0}\Phi(t,t_0) = -\Phi(t,t_0)A(t_0), \quad t \in I.$$

(2) Integrating the equation
$$\frac{\partial}{\partial_g s}\Phi(s,t_0) = A(s)\Phi(s,t_0), \quad s \in I,$$
in the sense of Stieltjes, from t to t_0, we get
$$\Phi(t,t_0) - \Phi(t_0,t_0) = \int_{t_0}^{t} A(s)\Phi(s,t_0)d_g s, \quad t \in I,$$
whereupon
$$\Phi(t,t_0) = \Phi(t_0,t_0) + \int_{t_0}^{t} A(s)\Phi(s,t_0)d_g s$$
$$= \mathscr{I} + \int_{t_0}^{t} A(s)\Phi(s,t_0)d_g s, \quad t \in I.$$

(3) We have
$$\Phi(t,t) - \Phi(t,t_0) = \int_{t_0}^{t} \frac{\partial}{\partial_g s}\Phi(t,s)d_g s, \quad t \in I.$$

From Item 1, we can write
$$\Phi(t,t) - \Phi(t,t_0) = -\int_{t_0}^{t} \Phi(t,s)A(s)d_g s, \quad t \in I,$$

i.e.,
$$\mathscr{I} - \Phi(t,t_0) = -\int_{t_0}^{t} \Phi(t,s)A(s)d_g s, \quad t \in I,$$

whereupon
$$\Phi(t,t_0) = \mathscr{I} + \int_{t_0}^{t} \Phi(t,s)A(s)d_g s, \quad t \in I.$$

This completes the proof. □

7.6 Stieltjes Adjoint Differential Systems

Now, suppose that Ψ is a fundamental matrix for the system (7.8). Then we differentiate the equation
$$\Psi(t)\Psi^{-1}(t) = \mathscr{I}, \quad t \in I,$$
in the sense of Stieltjes, and we get
$$\Psi(t)(\Psi^{-1})'_g(t) + \Psi'_g(t)\Psi^{-1}(t) = \mathscr{I}, \quad t \in I,$$
whereupon
$$\Psi(t)(\Psi^{-1})'_g(t) = -\Psi'_g(t)\Psi^{-1}(t), \quad t \in I,$$
or
$$\begin{aligned}(\Psi^{-1})'_g(t) &= -\Psi^{-1}(t)\Psi'_g(t)\Psi^{-1}(t)\\ &= -\Psi^{-1}(t)A(t)\Psi(t)\Psi^{-1}(t)\\ &= -\Psi^{-1}(t)A(t), \quad t \in I,\end{aligned}$$

and
$$((\Psi^{-1})^T)'_g(t) = -(A(t))^T (\Psi^{-1}(t))^T, \quad t \in I. \tag{7.21}$$

Definition 7.11. The Stieltjes differential system
$$x'_g = -(A(t))^T x, \quad t \in I, \tag{7.22}$$
is said to be the Stieltjes adjoint system to the system (7.8).

Theorem 7.21. *Let Ψ be a fundamental matrix of the system (7.8). Then, κ is a fundamental matrix to its Stieltjes adjoint system (7.22) if and only if*
$$(\kappa(x))^T \Psi(t) = C, \quad t \in I, \tag{7.23}$$
for some Stieltjes constant nonsingular $n \times n$-matrix C.

Proof. Suppose that Ψ is a fundamental matrix for the system (7.8). Then, in view of (7.21), it follows that $(\Psi^{-1})^T$ is a fundamental matrix of the system (7.22). If κ is a fundamental matrix of (7.22), applying Theorem 7.18, we obtain that there is a Stieltjes constant nonsingular $n \times n$-matrix C such that
$$\kappa(t) = (\Psi^{-1}(t))^T C, \quad t \in I, \tag{7.24}$$
whereupon
$$(\Psi(t))^T \kappa(t) = C^T, \quad t \in I, \tag{7.25}$$
which yields (7.23).

Conversely, Suppose that κ is a fundamental matrix of the system (7.22) that satisfies (7.23). Then, both (7.25) and (7.24) hold. Keeping in mind Theorem 7.18 and relation (7.24), we can say that κ is a fundamental matrix for the system (7.22). This completes the proof. □

7.7 The Method of Variation of Constants

Let $\Phi(\cdot, t_0)$ be the Stieltjes transitive matrix for the system (7.8). We will search a solution of the system
$$x'_g = A(t)x + f(t), \quad t \in I, \tag{7.26}$$

in the following form:
$$x(t) = \Phi(t,t_0)y(t), \quad t \in I,$$
for some function y defined on I. Since x is a solution to the system (7.26), we have
$$\begin{aligned}x'_g(t) &= A(t)x(t) + f(t) \\ &= A(t)\Phi(t,t_0)y(t) + f(t), \quad t \in I.\end{aligned} \quad (7.27)$$

Moreover,
$$\begin{aligned}x'_g(t) &= \Phi'_{tg}(t,t_0)y(t) + \Phi(t,t_0)y'_g(t) \\ &= A(t)\Phi(t,t_0)y(t) + \Phi(t,t_0)y'_g(t), \quad t \in I.\end{aligned}$$

Keeping in mind (7.27) and the last equation, we get
$$A(t)\Phi(t,t_0)y(t) + f(t) = A(t)\Phi(t,t_0)y(t) + \Phi(t,t_0)y'_g(t), \quad t \in I.$$

Hence,
$$\Phi(t,t_0)y'_g(t) = f(t), \quad t \in I,$$
and
$$\begin{aligned}y'_g(t) &= (\Phi(t,t_0))^{-1}f(t) \\ &= \Phi(t_0,t)f(t), \quad t \in I.\end{aligned}$$

Integrating the last equation, in the sense of Stieltjes, from t_0 to t, we find
$$\int_{t_0}^{t} y'_g(t)d_g s = \int_{t_0}^{t} \Phi(t_0,\tau)f(\tau)d_g\tau, \quad t \in I,$$
i.e.,
$$y(t) - y(t_0) = \int_{t_0}^{t} \Phi(t_0,\tau)f(\tau)d_g\tau, \quad t \in I,$$
i.e.,
$$y(t) = y(t_0) + \int_{t_0}^{t} \Phi(t_0,\tau)f(\tau)d_g\tau, \quad t \in I.$$

Note that

$$\begin{aligned}x(t_0) &= \Phi(t_0,t_0)y(t_0)\\ &= \mathscr{I}y(t_0)\\ &= y(t_0)\\ &= x_0.\end{aligned}$$

Thus,

$$y(t) = x_0 + \int_{t_0}^{t} \Phi(t_0,\tau)f(\tau)d_g\tau, \quad t \in I,$$

and

$$\begin{aligned}x(t) &= \Phi(t,t_0)\left(x_0 + \int_{t_0}^{t}\Phi(t_0,\tau)f(\tau)d_g\tau\right)\\ &= \Phi(t,t_0)x_0 + \int_{t_0}^{t}\Phi(t,t_0)\Phi(t_0,\tau)f(\tau)d_g\tau\\ &= \phi(t,t_0)x_0 + \int_{t_0}^{t}\Phi(t,\tau)f(\tau)d_g\tau, \quad t \in I,\end{aligned}$$

i.e.,

$$x(t) = \phi(t,t_0)x_0 + \int_{t_0}^{t}\Phi(t,\tau)f(\tau)d_g\tau, \quad t \in I. \qquad (7.28)$$

If Ψ is an arbitrary fundamental matrix, then there is a Stieltjes constant nonsingular matrix C so that

$$\Phi(t,t_0) = \Psi(t)C, \quad t \in I.$$

Hence,

$$\begin{aligned}\Phi(t_0,t) &= (\Phi(t,t_0))^{-1}\\ &= (\Psi(t)C)^{-1}\\ &= C^{-1}(\Psi(t))^{-1}, \quad t \in I.\end{aligned}$$

Then, applying (7.28), we find

$$x(t) = \Psi(t)Cx_0 + \int_{t_0}^{t} \Psi(t)CC^{-1}(\Psi(\tau))^{-1}d_g\tau$$

$$= \Psi(t)Cx_0 + \int_{t_0}^{t} \Psi(t)(\Psi(\tau))^{-1}d_g\tau, \quad t \in I.$$

Set

$$c = Cx_0.$$

Then

$$x(t) = \Psi(t)c + \int_{t_0}^{t} \Psi(t)(\Psi(\tau))^{-1}d_g\tau, \quad t \in I. \quad (7.29)$$

Exercise 7.6. Consider the Stieltjes differential system

$$x'_{1g} = x_1 + 2g(t),$$
$$x'_{2g} = x_1 + (g(t))^2, \quad t \in I.$$

(1) Find a fundamental system of solutions of the corresponding homogeneous system.
(2) Find the general solution of the given system.

7.8 Constant Coefficients

In this section, we suppose that A is an $n \times n$-matrix-valued constant function such that $A \in \mathscr{R}_g$. Consider the Stieltjes differential system

$$x'_g = Ax. \quad (7.30)$$

Theorem 7.22. *Let λ and ξ be an eigenpair of A. Then*

$$x(t) = e_{\lambda,g}(t, t_0)\xi, \quad t \in I,$$

is a solution of (7.30).

Proof. Since

$$A\xi = \lambda\xi,$$

for $t \in I$, we have

$$\begin{aligned} x'_g(t) &= (e_{\lambda,g})'_g(t,t_0)\xi \\ &= \lambda e_{\lambda,g}(t,t_0)\xi \\ &= e_{\lambda,g}(t,t_0)(\lambda\xi) \\ &= e_{\lambda,g}(t,t_0)A\xi \\ &= A\left(e_{\lambda,g}(t,t_0)\xi\right) \\ &= Ax(t), \quad t \in I. \end{aligned}$$

Thus,

$$x'_g(t) = Ax(t), \quad t \in I,$$

i.e., x is a solution of (7.30). This completes the proof. $\square$

Example 7.13. Consider the Stieltjes differential system

$$\begin{aligned} x'_{1g}(t) &= -3x_1 - 2x_2, \\ x'_{2g}(t) &= 3x_1 + 4x_2, \quad t \in I. \end{aligned}$$

Here

$$A = \begin{pmatrix} -3 & -2 \\ 3 & 4 \end{pmatrix}.$$

Then

$$\det(A - \lambda \mathscr{I}) = \begin{pmatrix} -3-\lambda & -2 \\ 3 & 4-\lambda \end{pmatrix}$$
$$= \lambda^2 - \lambda - 6,$$

whereupon $\det(A - \lambda \mathscr{I}) = 0$ yields $\lambda_1 = 3$ and $\lambda_2 = -2$. The matrix A is Stieltjes regressive provided $-2 \in \mathscr{R}_g$. Note that

$$\xi_1 = \begin{pmatrix} 1 \\ -3 \end{pmatrix} \quad \text{and} \quad \xi_2 = \begin{pmatrix} -2 \\ 1 \end{pmatrix}$$

are eigenvalues corresponding to λ_1 and λ_2, respectively. Therefore,

$$x(t) = c_1 e_{3,g}(t, t_0)\xi_1 + c_2 e_{-2,g}(t, t_0)\xi_2$$

$$= c_1 e_{3,g}(t, t_0) \begin{pmatrix} 1 \\ -3 \end{pmatrix} + c_2 e_{-2,g}(t, t_0) \begin{pmatrix} -2 \\ 1 \end{pmatrix}, \quad t \in I,$$

where c_1 and c_2 are real constants, is a general solution of the considered Stieltjes differential system provided $-2 \in \mathscr{R}_g$.

Example 7.14. Consider the Stieltjes differential system

$$x'_{1g}(t) = x_1(t) - x_2(t),$$
$$x'_{2g}(t) = -x_1(t) + 2x_2(t) - x_3(t),$$
$$x'_{3g}(t) = -x_2(t) + x_3(t), \quad t \in I.$$

Here

$$A = \begin{pmatrix} 1 & -1 & 0 \\ -1 & 2 & -1 \\ 0 & -1 & 1 \end{pmatrix}.$$

Then

$$\det(A - \lambda \mathscr{I}) = \det \begin{pmatrix} 1-\lambda & -1 & 0 \\ -1 & 2-\lambda & -1 \\ 0 & -1 & 1-\lambda \end{pmatrix}$$

$$= -(\lambda-1)^2(\lambda-2) + (\lambda-1) + (\lambda-1)$$
$$= (\lambda-1)\left(-(\lambda-1)(\lambda-2) + 2\right)$$
$$= (\lambda-1)\left(-\lambda^2 + 3\lambda\right)$$
$$= -\lambda(\lambda-1)(\lambda-3),$$

whereupon

$$\det(A - \lambda \mathscr{I}) = 0$$

yields $\lambda_1 = 0$, $\lambda_2 = 1$, and $\lambda_3 = 3$. The matrix A is Stieltjes regressive provided $1, 3 \in \mathscr{R}_g$. Note that

$$\xi_1 = \begin{pmatrix} 1 \\ 1 \\ 1 \end{pmatrix}, \quad \xi_2 = \begin{pmatrix} 1 \\ 0 \\ -1 \end{pmatrix}, \quad \text{and} \quad \xi_3 = \begin{pmatrix} 1 \\ -2 \\ 1 \end{pmatrix}$$

are eigenvalues corresponding to λ_1, λ_2, and λ_3, respectively. Consequently,

$$x(t) = c_1 \xi_1 + c_2 e_{1,g}(t, t_0) \xi_2 + c_3 e_{3,g}(t, t_0) \xi_3$$

$$= c_1 \begin{pmatrix} 1 \\ 1 \\ 1 \end{pmatrix} + c_2 e_{1,g}(t, t_0) \begin{pmatrix} 1 \\ 0 \\ -1 \end{pmatrix} + c_3 e_{3,g}(t, t_0) \begin{pmatrix} 1 \\ -2 \\ 1 \end{pmatrix}, \quad t \in I,$$

where c_1, c_2, and c_3 are constants, is a general solution of the considered Stieltjes differential system provided $1, 3 \in \mathscr{R}_g$.

Example 7.15. Consider the Stieltjes differential system

$$x'_{1g}(t) = -x_1(t) + x_2(t) + x_3(t),$$
$$x'_{2g}(t) = x_2(t) - x_3(t) + x_4(t),$$
$$x'_{3g}(t) = 2x_3(t) - 2x_4(t),$$
$$x'_{4g}(t) = 3x_4(t), \quad t \in I.$$

Here

$$A = \begin{pmatrix} -1 & 1 & 1 & 0 \\ 0 & 1 & -1 & 1 \\ 0 & 0 & 2 & -2 \\ 0 & 0 & 0 & 3 \end{pmatrix}.$$

Then

$$\det(A - \lambda \mathscr{I}) = \det \begin{pmatrix} -1-\lambda & 1 & 1 & 0 \\ 0 & 1-\lambda & -1 & 1 \\ 0 & 0 & 2-\lambda & -2 \\ 0 & 0 & 0 & 3-\lambda \end{pmatrix}$$

$$= (\lambda + 1)(\lambda - 1)(\lambda - 2)(\lambda - 3),$$

whereupon
$$\det(A - \lambda \mathscr{I}) = 0$$
yields $\lambda_1 = -1$, $\lambda_2 = 1$, $\lambda_3 = 2$, and $\lambda_4 = 3$. The matrix A is Stieltjes regressive provided $-1 \in \mathscr{R}_g$. Note that

$$\xi_1 = \begin{pmatrix} 1 \\ 0 \\ 0 \\ 0 \end{pmatrix}, \quad \xi_2 = \begin{pmatrix} 0 \\ 1 \\ 0 \\ 0 \end{pmatrix}, \quad \xi_3 = \begin{pmatrix} 0 \\ 0 \\ 1 \\ 0 \end{pmatrix}, \quad \text{and} \quad \xi_4 = \begin{pmatrix} 0 \\ 0 \\ 0 \\ 1 \end{pmatrix}$$

are eigenvectors corresponding to λ_1, λ_2, λ_3, and λ_4, respectively. Consequently,

$$x(t) = c_1 e_{-1,g}(t, t_0) \xi_1 + c_2 e_{1,g}(t, t_0) \xi_2 + c_3 e_{2,g}(t, t_0) \xi_3 + c_4 e_{5,g}(t, t_0) \xi_4$$

$$= c_1 e_{-1,g}(t, t_0) \begin{pmatrix} 1 \\ 0 \\ 0 \\ 0 \end{pmatrix} + c_2 e_{1,g}(t, t_0) \begin{pmatrix} 0 \\ 1 \\ 0 \\ 0 \end{pmatrix} + c_3 e_{2,g}(t, t_0) \begin{pmatrix} 0 \\ 0 \\ 1 \\ 0 \end{pmatrix}$$

$$+ c_4 e_{3,g}(t, t_0) \begin{pmatrix} 0 \\ 0 \\ 0 \\ 1 \end{pmatrix},$$

where $t \in I$ and c_1, c_2, c_3, and c_4 are real constants, is a general solution of the considered Stieltjes differential system.

Exercise 7.7. Find a general solution of the Stieltjes differential system
$$x'_{1g}(t) = x_2(t),$$
$$x'_{2g}(t) = x_1(t), \quad t \in I.$$

Answer 7.3.
$$x(t) = c_1 e_{1,g}(t, t_0) \begin{pmatrix} 1 \\ 1 \end{pmatrix} + c_2 e_{-1,g}(t, t_0) \begin{pmatrix} 1 \\ -1 \end{pmatrix}, \quad t \in I,$$

where $c_1, c_2 \in \mathbb{R}$, for which $-1 \in \mathscr{R}_g$.

Theorem 7.23. *Assume that $A \in \mathcal{R}_g$. If*
$$x(t) = u(t) + iv(t), \quad t \in I,$$
is a complex vector-valued solution of (7.30), *where u and v are real vector-valued functions on I, then u and v are real vector-valued solutions of* (7.30) *on I.*

Proof. We have
$$\begin{aligned} x'_g(t) &= A(t)x(t) \\ &= A(t)(u(t) + iv(t)) \\ &= A(t)u(t) + iA(t)v(t) \\ &= u'_g(t) + iv'_g(t), \quad t \in I. \end{aligned}$$

Equating real and imaginary parts, we get
$$u'_g(t) = A(t)u(t), \quad t \in I,$$
and
$$v'_g(t) = A(t)v(t), \quad t \in I.$$

This means that u and v are solutions of (7.30) on I. This completes the proof. □

Example 7.16. Consider the Stieltjes differential system
$$\begin{aligned} x'_{1g}(t) &= x_1(t) + x_2(t), \\ x'_{2g}(t) &= -x_1(t) + x_2(t), \quad t \in I. \end{aligned}$$

Here
$$A = \begin{pmatrix} 1 & 1 \\ -1 & 1 \end{pmatrix}.$$

Then
$$\det(A - \lambda \mathcal{I}) = \det \begin{pmatrix} 1 - \lambda & 1 \\ -1 & 1 - \lambda \end{pmatrix}$$
$$= (\lambda - 1)^2 + 1$$

$$= \lambda^2 - 2\lambda + 1 + 1$$
$$= \lambda^2 - 2\lambda + 2,$$

whereupon

$$\det(A - \lambda \mathscr{I}) = 0$$

yields

$$\lambda_{1,2} = 1 \pm i.$$

The matrix A is Stieltjes regressive provided $1 \pm i \in \mathscr{R}_g$. Note that

$$\xi = \begin{pmatrix} 1 \\ i \end{pmatrix}$$

is an eigenvector corresponding to the eigenvalue $\lambda = 1 + i$. We have

$$x(t) = e_{1+i,g}(t, t_0) \begin{pmatrix} 1 \\ i \end{pmatrix}$$

$$= e_{1,g}(t, t_0) \left(\cos_{\frac{1}{1+\Delta g(t)}, g}(t, t_0) + i \sin_{\frac{1}{1+\Delta g(t)}, g}(t, t_0) \right) \begin{pmatrix} 1 \\ i \end{pmatrix}$$

$$= e_{1,g}(t, t_0) \left(\begin{pmatrix} \cos_{\frac{1}{1+\Delta g(t)}, g}(t, t_0) \\ i \cos_{\frac{1}{1+\Delta g(t)}, g}(t, t_0) \end{pmatrix} + \begin{pmatrix} i \sin_{\frac{1}{1+\Delta g(t)}, g}(t, t_0) \\ -\sin_{\frac{1}{1+\Delta g(t)}, g}(t, t_0) \end{pmatrix} \right)$$

$$= e_{1,g}(t, t_0) \begin{pmatrix} \cos_{\frac{1}{1+\Delta g(t)}, g}(t, t_0) \\ -\sin_{\frac{1}{1+\Delta g(t)}, g}(t, t_0) \end{pmatrix}$$

$$+ i e_{1,g}(t, t_0) \begin{pmatrix} \sin_{\frac{1}{1+\Delta g(t)}, g}(t, t_0) \\ \cos_{\frac{1}{1+\Delta g(t)}, g}(t, t_0) \end{pmatrix}, \quad t \in I.$$

Consequently,

$$e_{1,g}(t, t_0) \begin{pmatrix} \cos_{\frac{1}{1+\Delta g(t)}, g}(t, t_0) \\ -\sin_{\frac{1}{1+\Delta g(t)}, g}(t, t_0) \end{pmatrix} \quad \text{and} \quad e_{1,g}(t, t_0) \begin{pmatrix} \sin_{\frac{1}{1+\Delta g(t)}, g}(t, t_0) \\ \cos_{\frac{1}{1+\Delta g(t)}, g}(t, t_0) \end{pmatrix},$$

$t \in I,$

are solutions of the considered Stieltjes differential system. Therefore,

$$x(t) = c_1 e_{1,g}(t,t_0) \begin{pmatrix} \cos_{\frac{1}{1+\Delta g(t)},g}(t,t_0) \\ -\sin_{\frac{1}{1+\Delta g(t)},g}(t,t_0) \end{pmatrix}$$
$$+ c_2 e_{1,g}(t,t_0) \begin{pmatrix} \sin_{\frac{1}{1+\Delta g(t)},g}(t,t_0) \\ \cos_{\frac{1}{1+\Delta g(t)},g}(t,t_0) \end{pmatrix}, \quad t \in I,$$

where $c_1, c_2 \in \mathbb{R}$, is a general solution of the considered Stieltjes differential system provided $1 \pm i \in \mathscr{R}_g$.

Example 7.17. Consider the Stieltjes differential system

$$x'_{1g}(t) = x_2(t),$$
$$x'_{2g}(t) = x_3(t),$$
$$x'_{3g}(t) = 2x_1(t) - 4x_2(t) + 3x_3(t), \quad t \in I.$$

Here

$$A = \begin{pmatrix} 0 & 1 & 0 \\ 0 & 0 & 1 \\ 2 & -4 & 3 \end{pmatrix}.$$

Then

$$\det(A - \lambda \mathscr{I}) = \det \begin{pmatrix} -\lambda & 1 & 0 \\ 0 & -\lambda & 1 \\ 2 & -4 & 3-\lambda \end{pmatrix}$$
$$= -\lambda^2(\lambda - 3) + 2 - 4\lambda$$
$$= -(\lambda^3 - 3\lambda^2 + 4\lambda - 2)$$
$$= -(\lambda - 1)(\lambda^2 - 2\lambda + 2),$$

whereupon

$$\det(A - \lambda \mathscr{I}) = 0$$

yields $\lambda_1 = 1$, $\lambda_{2,3} = 1 \pm i$. The matrix A is Stieltjes regressive provided $1, 1 \pm i \in \mathscr{R}_g$. Note that

$$\xi_1 = \begin{pmatrix} 1 \\ 1 \\ 1 \end{pmatrix} \quad \text{and} \quad \xi_2 \begin{pmatrix} 1 \\ 1+i \\ 2i \end{pmatrix}$$

are eigenvectors corresponding to the eigenvalues $\lambda_1 = 1$ and $\lambda_2 = 1 + i$, respectively. Further,

$$e_{1+i,g}(t,t_0) \begin{pmatrix} 1 \\ 1+i \\ 2i \end{pmatrix}$$

$$= e_{1,g}(t,t_0) \left(\cos_{\frac{1}{1+\Delta g(t)},g}(t,t_0) + i \sin_{\frac{1}{1+\Delta g(t)},g}(t,t_0) \right) \begin{pmatrix} 1 \\ 1+i \\ 2i \end{pmatrix}$$

$$= e_{1,g}(t,t_0) \left(\begin{pmatrix} \cos_{\frac{1}{1+\Delta g(t)},g}(t,t_0) \\ (1+i)\cos_{\frac{1}{1+\Delta g(t)},g}(t,t_0) \\ 2i\cos_{\frac{1}{1+\Delta g(t)},g}(t,t_0) \end{pmatrix} + i \begin{pmatrix} \sin_{\frac{1}{1+\Delta g(t)},g}(t,t_0) \\ (1+i)\sin_{\frac{1}{1+\Delta g(t)},g}(t,t_0) \\ 2i\sin_{\frac{1}{1+\Delta g(t)},g}(t,t_0) \end{pmatrix} \right)$$

$$= e_{1,g}(t,t_0) \left(\begin{pmatrix} \cos_{\frac{1}{1+\Delta g(t)},g}(t,t_0) \\ (1+i)\cos_{\frac{1}{1+\Delta g(t)},g}(t,t_0) \\ 2i\cos_{\frac{1}{1+\Delta g(t)},g}(t,t_0) \end{pmatrix} + \begin{pmatrix} i\sin_{\frac{1}{1+\Delta g(t)},g}(t,t_0) \\ (-1+i)\sin_{\frac{1}{1+\Delta g(t)},g}(t,t_0) \\ -2\sin_{\frac{1}{1+\Delta g(t)},g}(t,t_0) \end{pmatrix} \right)$$

$$= e_{1,g}(t,t_0) \left(\begin{pmatrix} \cos_{\frac{1}{1+\Delta g(t)},g}(t,t_0) \\ \cos_{\frac{1}{1+\Delta g(t)},g}(t,t_0) - \sin_{\frac{1}{1+\Delta g(t)},g}(t,t_0) \\ -2\sin_{\frac{1}{1+\Delta g(t)},g}(t,t_0) \end{pmatrix} \right.$$

$$\left. + i \begin{pmatrix} \sin_{\frac{1}{1+\Delta g(t)},g}(t,t_0) \\ \cos_{\frac{1}{1+\Delta g(t)},g}(t,t_0) + \sin_{\frac{1}{1+\Delta g(t)},g}(t,t_0) \\ 2\cos_{\frac{1}{1+\Delta g(t)},g}(t,t_0) \end{pmatrix} \right), \quad t \in I.$$

Consequently,

$$x(t) = e_{1,g}(t,t_0)\left(c_1\begin{pmatrix}1\\1\\1\end{pmatrix} + c_2\begin{pmatrix}\cos_{\frac{1}{1+\Delta g(t)},g}(t,t_0)\\ \cos_{\frac{1}{1+\Delta g(t)},g}(t,t_0) - \sin_{\frac{1}{1+\Delta g(t)},g}(t,t_0)\\ -2\sin_{\frac{1}{1+\Delta g(t)},g}(t,t_0)\end{pmatrix}\right.$$

$$\left.+c_3\begin{pmatrix}\sin_{\frac{1}{1+\Delta g(t)},g}(t,t_0)\\ \cos_{\frac{1}{1+\Delta g(t)},g}(t,t_0) + \sin_{\frac{1}{1+\Delta g(t)},g}(t,t_0)\\ 2\cos_{\frac{1}{1+\Delta g(t)},g}(t,t_0)\end{pmatrix}\right), \quad t \in I,$$

where $c_1, c_2, c_3 \in \mathbb{R}$, is a general solution of the considered Stieltjes differential system provided $1, 1 \pm i \in \mathscr{R}_g$.

Exercise 7.8. Find a general solution of the Stieltjes differential system

$$x'_{1g}(t) = x_1(t) - 2x_2(t) + x_3(t),$$
$$x'_{2g}(t) = -x_1(t) + x_3(t),$$
$$x'_{3g}(t) = x_1(t) - 2x_2(t) + x_3(t), \quad t \in I.$$

Theorem 7.24 (The Stieltjes–Putzer Algorithm). *Let $A \in \mathscr{R}_g$ be a constant $n \times n$-matrix and $\lambda_1, \lambda_2, \ldots, \lambda_n$ be the eigenvalues of A. Then*

$$e_{A,g}(t,t_0) = \sum_{k=0}^{n-1} r_{k+1}(t) P_k, \quad t \in I,$$

where

$$r(t) = \begin{pmatrix} r_1(t) \\ \vdots \\ r_n(t) \end{pmatrix}, \quad t \in I,$$

is the solution of the SIVP

$$r'_g = \begin{pmatrix} \lambda_1 & 0 & 0 & \cdots & 0 \\ 1 & \lambda_2 & 0 & \cdots & 0 \\ 0 & 1 & \lambda_3 & \cdots & 0 \\ \vdots & \vdots & \vdots & \vdots & \vdots \\ 0 & 0 & 0 & \cdots & \lambda_n \end{pmatrix} r, \quad r(t_0) = \begin{pmatrix} 1 \\ 0 \\ 0 \\ \vdots \\ 0 \end{pmatrix}, \quad (7.31)$$

and the P-matrices are recursively defined by

$$P_0 = \mathscr{I}$$

and

$$P_{k+1} = (A - \lambda_{k+1}\mathscr{I})P_k, \quad 0 \leq k \leq n-1.$$

Proof. Since A is Stieltjes regressive, in view of Theorem 7.5, all eigenvalues of A are Stieltjes regressive. Hence, the SIVP (7.31) has a unique solution. We set

$$X(t) = \sum_{k=0}^{n-1} r_{k+1}(t) P_k, \quad t \in I. \quad (7.32)$$

We have

$$P_1 = (A - \lambda_1 \mathscr{I}) P_0$$
$$= (A - \lambda_1 \mathscr{I}),$$
$$P_2 = (A - \lambda_2 \mathscr{I}) P_1$$
$$= (A - \lambda_2 I)(A - \lambda_1 \mathscr{I}),$$
$$\vdots$$
$$P_n = (A - \lambda_n \mathscr{I}) P_{n-1}$$
$$= (A - \lambda_n \mathscr{I}) \ldots (A - \lambda_1 \mathscr{I})$$
$$= 0.$$

Therefore
$$X'_g(t) = \sum_{k=0}^{n-1} D_g r_{k+1}(t) P_k, \quad t \in I,$$

and

$X'_g(t) - AX(t)$

$$= \sum_{k=0}^{n-1} r'_{k+1,g}(t) P_k - A \sum_{k=0}^{n-1} r_{k+1}(t) P_k$$

$$= r'_{1g}(t) P_0 + \sum_{k=1}^{n-1} r'_{k+1,g}(t) P_k - A \sum_{k=0}^{n-1} r_{k+1}(t) P_k$$

$$= \lambda_1 r_1(t) P_0 + \sum_{k=1}^{n-1} (r_k(t) + \lambda_{k+1} r_{k+1}(t)) P_k - \sum_{k=1}^{n-1} r_{k+1}(t) A P_k$$

$$= \sum_{k=1}^{n-1} r_k(t) P_k + \lambda_1 r_1(t) P_0 + \sum_{k=1}^{n-1} \lambda_{k+1} r_{k+1}(t) P_k - \sum_{k=0}^{n-1} r_{k+1}(t) A P_k$$

$$= \sum_{k=1}^{n-1} r_k(t) P_k + \sum_{k=0}^{n-1} \lambda_{k+1} r_{k+1}(t) P_k - \sum_{k=0}^{n-1} r_{k+1}(t) A P_k$$

$$= \sum_{k=1}^{n-1} r_k(t) P_k - \sum_{k=0}^{n-1} (A - \lambda_{k+1} I) r_{k+1}(t) P_k$$

$$= \sum_{k=1}^{n-1} r_k(t) P_k - \sum_{k=0}^{n-1} r_{k+1}(t) P_{k+1}$$

$$= -r_n(t) P_n$$

$$= 0, \quad t \in I.$$

Thus,
$$X'_g(t) = AX(t), \quad t \in I.$$

Also,
$$X(t_0) = \sum_{k=0}^{n-1} r_{k+1}(t_0) P_k$$
$$= r_1(t_0) P_0$$
$$= \mathscr{I}.$$

Hence, X defined in (7.32) is really a solution of the SIVP (7.31). This completes the proof. □

Example 7.18. Consider the Stieltjes differential system
$$x'_{1g}(t) = 2x_1(t) + x_2(t) + 2x_3(t),$$
$$x'_{2g}(t) = 4x_1(t) + 2x_2(t) + 4x_3(t),$$
$$x'_{3g}(t) = 2x_1(t) + x_2(t) + 2x_3(t), \quad t \in I.$$

Here
$$A = \begin{pmatrix} 2 & 1 & 2 \\ 4 & 2 & 4 \\ 2 & 1 & 2 \end{pmatrix}.$$

Then
$$\det(A - \lambda \mathscr{I}) = \det \begin{pmatrix} 2-\lambda & 1 & 2 \\ 4 & 2-\lambda & 4 \\ 2 & 1 & 2-\lambda \end{pmatrix}$$
$$= -(\lambda - 2)^3 + 8 + 8 + 4(\lambda - 2) + 4(\lambda - 2) + 4(\lambda - 2)$$
$$= -(\lambda - 2)^3 + 12(\lambda - 2) + 16$$
$$= -\left(\lambda^3 - 6\lambda^2 + 12\lambda - 8 - 12\lambda + 24 - 16\right)$$
$$= -\left(\lambda^3 - 6\lambda^2\right)$$
$$= -\lambda^2(\lambda - 6),$$

whereupon
$$\det(A - \lambda \mathscr{I}) = 0$$

yields $\lambda_1 = \lambda_2 = 0$ and $\lambda_3 = 6$. Note that matrix A is Stieltjes regressive provided $6 \in \mathscr{R}_g$. Consider the following SIVPs:

$$r'_{1g}(t) = 0, \quad r_1(t_0) = 1,$$
$$r'_{2g}(t) = r_1(t), \quad r_2(t_0) = 0,$$
$$r'_{3g}(t) = r_2(t) + 6r_3(t), \quad r_3(t_0) = 0, \quad t \in I.$$

Then, we have

$$r_1(t) = 1, \quad t \in I,$$
$$r'_{2g}(t) = 1, \quad r_2(t_0) = 0, \quad t \in I.$$

Thus,

$$r_2(t) = g(t) - g(t_0), \quad t \in I,$$

and

$$r'_{3g}(t) = g(t) - g(t_0) + 6r_3(t), \quad r_3(t_0) = 0, \quad t \in I.$$

This gives

$$r_3(t) = \int_{s_0}^{t} e_{\ominus_g 6, g}(\tau, t)(g(\tau) - g(t_0)) d_g \tau, \quad t \in I.$$

Next, we have

$$P_0 = \begin{pmatrix} 1 & 0 & 0 \\ 0 & 1 & 0 \\ 0 & 0 & 1 \end{pmatrix},$$

$$P_1 = (A - \lambda_1 \mathscr{I}) P_0$$
$$= A P_0$$
$$= A$$
$$= \begin{pmatrix} 2 & 1 & 2 \\ 4 & 2 & 4 \\ 2 & 1 & 2 \end{pmatrix},$$

$$P_2 = (A - \lambda_1 I)(A - \lambda_2 I)$$
$$= A^2 I$$
$$= A^2$$
$$= \begin{pmatrix} 2 & 1 & 2 \\ 4 & 2 & 4 \\ 2 & 1 & 2 \end{pmatrix} \begin{pmatrix} 2 & 1 & 2 \\ 4 & 2 & 4 \\ 2 & 1 & 2 \end{pmatrix}$$
$$= \begin{pmatrix} 12 & 6 & 12 \\ 24 & 12 & 24 \\ 12 & 6 & 12 \end{pmatrix},$$

and

$$P_3 = 0.$$

Hence, keeping in mind the Stieltjes–Putzer algorithm given in Theorem 7.24, we obtain

$$e_{A,g}(t,t_0) = r_1(t)P_0 + r_2(t)P_1 + r_3(t)P_2$$
$$= \begin{pmatrix} 1 & 0 & 0 \\ 0 & 1 & 0 \\ 0 & 0 & 1 \end{pmatrix} + (g(t) - g(t_0)) \begin{pmatrix} 2 & 1 & 2 \\ 4 & 2 & 4 \\ 2 & 1 & 2 \end{pmatrix}$$
$$+ \left(\int_{t_0}^t e_{\ominus_g 6, g}(\tau, t)(g(\tau) - g(t_0)) d_g \tau \right) \begin{pmatrix} 12 & 6 & 12 \\ 24 & 12 & 24 \\ 12 & 6 & 12 \end{pmatrix},$$
$$t \in I,$$

and

$$\begin{pmatrix} x_1(t) \\ x_2(t) \\ x_3(t) \end{pmatrix} = e_{A,g}(t, t_0) \begin{pmatrix} c_1 \\ c_2 \\ c_3 \end{pmatrix}, \quad t \in I,$$

where $c_1, c_2, c_3 \in \mathbb{R}$, is a general solution of the considered Stieltjes differential system provided $6 \in \mathscr{R}_g$.

Exercise 7.9. Using the Stieltjes–Putzer algorithm, find $e_{A,g}(t,t_0)$, where

(1)
$$A = \begin{pmatrix} 1 & 2 \\ -1 & 3 \end{pmatrix},$$

(2)
$$A = \begin{pmatrix} 1 & -1 & 1 \\ 1 & 0 & 2 \\ -1 & 1 & 1 \end{pmatrix}.$$

7.9 Nonlinear Stieltjes Differential Systems

In this section, we investigate the following nonlinear system:
$$x'_{ig} = f_i(t,x), \quad i \in \{1,\ldots,n\}, \quad t \in [a,b], \tag{7.33}$$
subject to the initial condition
$$x_i(a) = x_{0i}, \quad i \in \{1,\ldots,n\}, \tag{7.34}$$
where

(A1) $[a,b] \subset I$ and $s_0 \in [a,b]$,
(A2) $f := (f_1,\ldots,f_n)$, $f_i \in \mathscr{C}([a,b] \times \mathbb{R})$, $i \in \{1,\ldots,n\}$, $x_0 = (x_{01},\ldots,x_{0n}) \in \mathbb{R}^n$, and $x = (x_1,\ldots,x_n)$ is the unknown function.

Denote $X = (\mathscr{C}([a,b]))^n$. In $\mathscr{C}([a,b] \times \mathbb{R})$, we define a norm
$$\|u\|_1 = \sup_{(t,y) \in [a,b] \times \mathbb{R}} |u(t,y)|,$$
provided it exists. In X, we define a norm
$$\|x\| = \max_{i \in \{1,\ldots,n\}} \|x_i\|_1.$$

In this section, we start our investigations with the following useful lemma.

Lemma 7.1. *Assume that (A1) and (A2) hold. Then, $x \in (\mathscr{C}_g([a,b]))^n$ is a solution to the SIVP (7.33)–(7.34) if and only if $x_i \in \mathscr{C}_g^1([a,b])$, $i \in \{1,\ldots,n\}$, satisfies*

$$x_i(t) = x_{0i} + \int_a^t f_i(s, x(s)) d_g s, \quad t \in [a,b], \quad i \in \{1,\ldots,n\}. \quad (7.35)$$

Proof. Assume that $x_i \in \mathscr{C}_g^1([a,b])$, $i \in \{1,\ldots,n\}$, satisfies (7.35). Since $f_i \in \mathscr{C}([a,b] \times \mathbb{R})$, $i \in \{1,\ldots,n\}$, we have that

$$\int_a^t f_i(s, x(s)) d_g s \in \mathscr{C}_g^1([a,b]), \quad i \in \{1,\ldots,n\}.$$

Thus, $x_i \in \mathscr{C}_g^1([a,b])$, $i \in \{1,\ldots,n\}$. Now, differentiating (7.35) in the sense of Stieltjes, we get (7.33). Putting $t = a$ into (7.35), we obtain (7.34). Thus, x is a solution to the SIVP (7.33)–(7.34). Conversely, assume that x is a solution to the SIVP (7.33)–(7.34). Now, integrating (7.33) in the sense of Stieltjes and using the initial condition (7.34), we obtain (7.35). This completes the proof. □

In addition, suppose the following:

(A3)
$$|f_i(t, x_1) - f_i(t, x_2)| \le A \sum_{l=1}^n |x_{1l} - x_{2l}|, \quad t \in [a,b], \quad x_1, x_2 \in \mathbb{R}^n,$$

$i \in \{1,\ldots,n\}$, for some positive constant A.
(A4) $|f_i(t, x)| \le M$, $t \in [a,b]$, $x \in \mathbb{R}^n$, $i \in \{1,\ldots,n\}$, for some nonnegative constant M.

Theorem 7.25. *Assume that (A1)–(A4) hold. Then the SIVP (7.33)–(7.34) has a unique solution $x \in (\mathscr{C}_g([a,b]))^n$.*

Proof. We define the sequence

$$x_{i0}(t) = x_{i0},$$

$$x_{il}(t) = x_{i0}(t) + \int_a^t f(s, x_{l-1}(s)) d_g s, \quad t \in [a,b], \quad l \in \mathbb{N}.$$

We have
$$|x_{i1}(t) - x_{i0}(t)| = \left|\int_a^t f_i(s, x_0(s))d_g s\right|$$
$$\leq \int_a^t |f_i(s, x_0(s))|d_g s$$
$$\leq M \int_a^t d_g s$$
$$= M h_1(t, a), \quad t \in [a, b].$$

Thus,
$$|x_{i1}(t) - x_{i0}(t)| \leq M h_1(t, a), \quad t \in [a, b], \quad i \in \{1, \ldots, n\}.$$

Assume that
$$|x_{il-1}(t) - x_{il-2}(t)|$$
$$\leq M n^{l-2} A^{l-2} h_{l-1}(t, a), \quad t \in [a, b], \quad i \in \{1, \ldots, n\}, \quad (7.36)$$
for some $l \in \mathbb{N}$, $l \geq 2$. We will prove that
$$|x_{il}(t) - x_{il-1}(t)| \leq M n^{l-1} A^{l-1} h_l(t, a), \quad t \in [a, b], \quad i \in \{1, \ldots, n\}.$$

Actually, we have
$$|x_{il}(t) - x_{il-1}(t)| = \left|\int_a^t (f_i(s, x_l(s)) - f_i(s, x_{l-1}(s)))\, d_g s\right|$$
$$\leq \int_a^t |f_i(s, x_l(s)) - f_i(s, x_{l-1}(s))|\, d_g s$$
$$\leq A \sum_{m=1}^n \int_a^t |x_{ml}(s) - x_{ml-1}(s)|d_g s$$
$$\leq A M n^{l-2} A^{l-2} \sum_{m=1}^n \int_a^t h_{l-1}(s, a) d_g s$$
$$= M A^{l-1} n^{l-1} h_l(t, a), \quad t \in [a, b],$$

i.e.,
$$|x_{il}(t) - x_{il-1}(t)| \leq MA^{l-1}n^{l-1}h_l(t,a), \quad t \in [a,b], \quad i \in \{1,\ldots,n\}.$$

Thus, (7.36) holds for $l \in \mathbb{N}$, $l \geq 2$. Note that

$$\left|\lim_{l\to\infty} (x_{il}(t) - x_{i0}(t))\right| = \left|\sum_{l=1}^{\infty} (x_{il}(t) - x_{il-1}(t))\right|$$

$$\leq \sum_{l=1}^{\infty} |x_{il}(t) - x_{il-1}(t)|$$

$$\leq M \sum_{l=1}^{\infty} A^{l-1}n^{l-1}h_l(t,a)$$

$$= \frac{M}{An} \sum_{l=1}^{\infty} A^l n^l h_l(t,a)$$

$$\leq \frac{M}{An} \sum_{l=1}^{\infty} A^l n^l h_l(b,a)$$

$$= \frac{M}{An} e_{An,g}(b,a), \quad t \in [a,b].$$

Thus,
$$\left|\lim_{l\to\infty} (x_{il}(t) - x_{i0}(t))\right| < \infty, \quad i \in \{1,\ldots,n\}, \quad t \in [a,b].$$

So, the series
$$\sum_{l=1}^{\infty} (x_{il}(t) - x_{il-1}(t)), \quad i \in \{1,\ldots,n\},$$

is uniformly convergent on $[a,b]$. Hence,
$$\lim_{l\to\infty} (x_{il}(t) - x_{i0}(t))$$
$$= \sum_{l=1}^{\infty} (x_{il}(t) - x_{il-1}(t)), \quad i \in \{1,\ldots,n\}, \quad t \in [a,b],$$

and then,
$$x(t) = \lim_{l \to \infty} x_l(t), \quad t \in [a, b].$$

Further, $x_i(t)$, $t \in [a, b]$, $i \in \{1, \ldots, n\}$, satisfy (7.35). From here and Lemma 7.1, we conclude that $x(t)$, $t \in [a, b]$, is a solution to the SIVP (7.33)–(7.34). Next, assume that the SIVP (7.33)–(7.34) has another solution $y(t)$, $t \in [a, b]$. By Lemma 7.1, we have that

$$y_i(t) = x_{i0}(t) + \int_a^t f_i(s, y(s)) d_g s, \quad t \in [a, b], \quad i \in \{1, \ldots, n\}.$$

Note that
$$|y_i(t) - x_{i0}(t)| = \left| \int_a^t f_i(s, y(s)) d_g s \right|$$
$$\leq \int_a^t |f_i(s, y(s))| d_g s$$
$$\leq M \int_a^t d_g s$$
$$= M h_1(t, a), \quad t \in [a, b].$$

Thus,
$$|y_i(t) - x_{i0}(t)| \leq M h_1(t, a), \quad t \in [a, b], \quad i \in \{1, \ldots, n\}.$$

Assume that
$$|x_{il-1}(t) - y_i(t)|$$
$$\leq M n^{l-2} A^{l-2} h_{l-1}(t, a), \quad t \in [a, b], \quad i \in \{1, \ldots, n\}, \quad (7.37)$$

for some $l \in \mathbb{N}$, $l \geq 2$. We will prove that
$$|x_{il}(t) - y_i(t)| \leq M n^{l-1} A^{l-1} h_l(t, a), \quad t \in [a, b], \quad i \in \{1, \ldots, n\}.$$

In fact, we have
$$|x_{il}(t) - y_i(t)| = \left| \int_a^t (f_i(s, x_{l-1}(s)) - f_i(s, y(s))) d_g s \right|$$
$$\leq \int_a^t |f_i(s, y(s)) - f_i(s, x_{l-1}(s))| d_g s$$

$$\leq A \sum_{m=1}^{n} \int_{a}^{t} |y_m(s) - x_{ml-1}(s)| d_g s$$

$$\leq A M n^{l-2} A^{l-2} \sum_{m=1}^{n} \int_{a}^{t} h_{l-1}(s,a) d_g s$$

$$= M A^{l-1} n^{l-1} h_l(t,a), \quad t \in [a,b],$$

i.e.,

$$|x_{il}(t) - y_i(t)| \leq M A^{l-1} n^{l-1} h_l(t,a), \quad t \in [a,b], \quad i \in \{1, \ldots, n\}.$$

Thus, (7.37) holds for $l \in \mathbb{N}$, $l \geq 2$. Since

$$\lim_{l \to \infty} M A^l n^l h_l(t,a) = 0, \quad t \in [a,b],$$

we conclude that

$$x_i(t) - y_i(t) = \lim_{l \to \infty} (x_{il}(t) - y_i(t))$$

$$= 0, \quad t \in [a,b], \quad i \in \{1, \ldots, n\}.$$

Hence, the SIVP (7.33)–(7.34) has unique solution in $(\mathscr{C}_g([a,b]))^n$. This completes the proof. □

7.10 Advanced Practical Problems

Problem 7.1. Let $I = \mathbb{R}$. Define $g(t) = 3t$ and

$$A(t) = \begin{pmatrix} \frac{t^2+2}{t+1} & t^2 + 3t \\ 4t - 1 & 3t \end{pmatrix}, \quad t \in I.$$

Then find $A'_g(t)$, $t \in I$.

Answer 7.4.

$$A'_g(t) = \begin{pmatrix} \frac{t^2+2t-1}{3(t+1)^2} & \frac{2t+3}{3} \\ \frac{4}{3} & 1 \end{pmatrix}, \quad t \in I.$$

Problem 7.2. Let $I = \mathbb{R}$. Define $g(t) = 3t$,

$$A(t) = \begin{pmatrix} t+10 & t^2-2t+2 \\ t & t^2+t+1 \end{pmatrix}, \quad \text{and} \quad B(t) = \begin{pmatrix} t-2 & t+1 \\ t^2 & t^3-1 \end{pmatrix}, \quad t \in I.$$

Then show that

$$(AB)'_g(t) = A'_g(t)B(t) + A(t)B'_g(t), \quad t \in I.$$

Problem 7.3. Let $I = \mathbb{R}$. Define $g(t) = 3t$,

$$A(t) = \begin{pmatrix} 1 & -1 \\ 0 & 1 \end{pmatrix}, \quad \text{and} \quad B(t) = \begin{pmatrix} 2 & 0 \\ -1 & 1 \end{pmatrix}, \quad t \in I.$$

Then find

$$(A \oplus_g B)(t), \quad t \in I.$$

Answer 7.5.

$$\begin{pmatrix} 3(t+2\sqrt{t}+2) & -(t+2\sqrt{t}+2) \\ -(t+2\sqrt{t}+2) & t+2\sqrt{t}+3 \end{pmatrix}, \quad t \in I.$$

Problem 7.4. Find a general solution of the Stieltjes differential system

$$x'_{1g}(t) = 2x_1(t) + 3x_2(t),$$
$$x'_{2g}(t) = x_1(t) + 4x_2(t), \quad t \in I.$$

Answer 7.6.

$$x(t) = c_1 e_{1,g}(t, t_0) \begin{pmatrix} 3 \\ -1 \end{pmatrix} + c_2 e_{5,g}(t, t_0) \begin{pmatrix} 1 \\ 1 \end{pmatrix}, \quad t \in I,$$

where $c_1, c_2 \in \mathbb{R}$, provided $1, 5 \in \mathscr{R}_g$.

Problem 7.5. Find a general solution of the Stieltjes differential system

$$x'_{1g}(t) = -x_1(t) - x_2(t) - x_3(t),$$
$$x'_{2g}(t) = x_1(t) - x_2(t) + 3x_3(t),$$
$$x'_{3g}(t) = x_1(t) - x_2(t) + 4x_3(t), \quad t \in I.$$

Problem 7.6. Using the Stieltjes–Putzer algorithm, find $e_{A,g}(t,t_0)$, where

(1)
$$A = \begin{pmatrix} 2 & 3 \\ 1 & -4 \end{pmatrix},$$

(2)
$$A = \begin{pmatrix} -1 & 2 & 3 \\ 1 & 1 & -4 \\ 1 & -1 & 2 \end{pmatrix}.$$

Chapter 8

Qualitative Analysis of Stieltjes Differential Systems

In this chapter, we study the qualitative properties of Stieltjes differential systems. First, we present the periodic solution and its properties, and then, we discuss asymptotic behaviour of solutions of Stieltjes differential systems.

Suppose that $I \subseteq \mathbb{R}$ and $g\colon I \to \mathbb{R}$ is a monotone nondecreasing function that is continuous from the left everywhere. Let ω be a positive real number. An ω-periodic function $x\colon I \to \mathbb{R}$ is a function defined on I with the property:
$$x(t) = x(t+\omega), \quad t \in I.$$
If an ω-periodic function x satisfy the Stieltjes differential system, then we say that x is a ω-periodic solution of that Stieltjes differential system.

8.1 Linear Periodic Stieltjes Differential Systems

In this section, we investigate the Stieltjes differential system
$$x'_g = A(t)x + f(t), \quad t \in I, \tag{8.1}$$
where A is an $n \times n$-matrix with entries $a_{ij} \in \mathscr{C}(I)$, $i,j \in \{1, \ldots, n\}$, which are ω-periodic, and $f \in \mathscr{C}(I)$ is a ω-periodic function. The corresponding homogeneous system is
$$x'_g = A(t)x, \quad t \in I. \tag{8.2}$$

Theorem 8.1. *The system* (8.1) *has a nontrivial ω-periodic solution x if and only if*

$$x(0) = x(\omega). \tag{8.3}$$

Proof. Assume that x is a nontrivial ω-periodic solution of (8.1). Then by the definition for a periodic function, we get (8.3). Conversely, suppose that (8.3) holds. Set

$$v(t) = x(t + \omega), \quad t \in I.$$

Then,

$$\begin{aligned} v'_g(t) &= x'_g(t + \omega) \\ &= A(t + \omega)x(t + \omega) + f(t + \omega) \\ &= A(t)v(t) + f(t), \quad t \in I, \end{aligned}$$

i.e.,

$$v'_g(t) = A(t)v(t) + f(t), \quad t \in I.$$

Thus, v is a solution to the system (8.1) for which

$$\begin{aligned} v(0) &= x(\omega) \\ &= x(0). \end{aligned}$$

Since the IVP for (8.1) has a unique solution, we can conclude that

$$v(t) = x(t), \quad t \in I,$$

i.e.,

$$x(t) = x(t + \omega), \quad t \in I.$$

This completes the proof. □

Corollary 8.1. *Let Φ be the Stieltjes transitive matrix for system* (8.2). *Then, the system* (8.2) *has a nontrivial ω-periodic solution if and only if*

$$\det(\Phi(0) - \Phi(\omega)) = 0. \tag{8.4}$$

Proof. Since Φ is the Stieltjes transitive matrix of (8.2), then any solution of (8.2) has the form

$$x(t) = \Phi(t)c, \quad t \in I,$$

where c is a Stieltjes constant vector. By Theorem 8.1, it follows that x is a nontrivial ω-periodic solution of (8.2) if and only if

$$x(0) = x(\omega),$$

equivalently,

$$\Phi(0)c = \Phi(\omega)c,$$

equivalently,

$$(\Phi(0) - \Phi(\omega))c = 0,$$

equivalently, (8.4) holds. This completes the proof. □

Corollary 8.2. *The system*

$$x'_g = Bx, \tag{8.5}$$

where B is a constant matrix, has a nontrivial ω-periodic solution if and only if

$$\det(\mathscr{I} - B\omega) = 0. \tag{8.6}$$

Proof. Note that

$$Bt, \quad t \in I,$$

is the Stieltjes transitive matrix for (8.5). Hence, applying Corollary 8.1, we get that the system (8.5) has a nontrivial ω-periodic solution if and only if (8.6) holds. This completes the proof. □

Example 8.1. Let

$$B = \begin{pmatrix} 0 & -\frac{2}{3} & -1 \\ 0 & -\frac{1}{3} & -\frac{2}{3} \\ -\frac{1}{3} & -\frac{4}{3} & -\frac{4}{3} \end{pmatrix}.$$

Then

$$3B = \begin{pmatrix} 0 & -2 & -3 \\ 0 & -1 & -2 \\ -1 & -4 & -4 \end{pmatrix},$$

$$\mathscr{I} - 3B = \begin{pmatrix} 1 & 0 & 0 \\ 0 & 1 & 0 \\ 0 & 0 & 1 \end{pmatrix} \begin{pmatrix} 0 & -2 & -3 \\ 0 & -1 & -2 \\ -1 & -4 & -4 \end{pmatrix}$$

$$= \begin{pmatrix} 1 & 2 & 3 \\ 0 & 2 & 2 \\ 1 & 4 & 5 \end{pmatrix}$$

and

$$\det(\mathscr{I} - 3B) = \det \begin{pmatrix} 1 & 2 & 3 \\ 0 & 2 & 2 \\ 1 & 4 & 5 \end{pmatrix}$$

$$= 10 + 4 - 6 - 8$$

$$= 0.$$

Thus, the system (8.5) has at least one nontrivial 3-periodic solution.

Corollary 8.3. *The system (8.1) has a unique nontrivial ω-periodic solution if and only if the system (8.2) has no nontrivial ω-periodic solution.*

Proof. Let Φ be the Stieltjes transitive matrix for (8.2) and let x be a nontrivial ω-periodic solution of (8.1). Then, this x is given by

$$x(t) = \Phi(t)c + \int_0^t \Phi(t)(\Phi(\tau))^{-1} f(\tau) d_g \tau, \quad t \in I.$$

Now, in view of Theorem 8.1, it follows that system (8.1) has a nontrivial ω-periodic solution if and only if

$$\Phi(0)c = \Phi(\omega)c + \int_0^\omega \Phi(\omega)(\Phi(\tau))^{-1} f(\tau) d_g \tau,$$

equivalently,

$$(\Phi(0) - \Phi(\omega))c = \int_0^t \Phi(t)(\Phi(\tau))^{-1} f(\tau) d_g \tau,$$

equivalently,

$$\det(\Phi(0) - \Phi(\omega)) \neq 0,$$

equivalently, the system (8.2) has no nontrivial ω-periodic solutions. This completes the proof. □

Theorem 8.2. *Let $\Psi(t)$, $t \in I$, be a fundamental matrix for the system (8.2). Then, $\Psi(t+\omega)$, $t \in I$, is also a fundamental matrix for (8.2).*

Proof. We have
$$\Psi'_g(t) = A(t)\Psi(t), \quad t \in I,$$
and
$$\det \Psi(t) \neq 0, \quad t \in I$$
Hence,
$$\Psi'_g(t+\omega) = A(t+\omega)\Psi(t+\omega)$$
$$= A(t)\Psi(t+\omega), \quad t \in I,$$
and
$$\det \Psi(t+\omega) \neq 0, \quad t \in I.$$
Thus, $\Psi(t+\omega)$ is a fundamental matrix for (8.2). This completes the proof. □

Theorem 8.3. *Let Ψ be a fundamental matrix for the system (8.2). Then there exists a ω-periodic Stieltjes nonsingular matrix P and a Stieltjes constant nonsingular matrix R such that*
$$\Psi(t) = P(t)Rt, \quad t \in I.$$

Proof. Since Ψ be a fundamental matrix for the system (8.2), in view of Theorem 8.2, it follows that $\Psi(t+\omega)$, $t \in I$, is also fundamental matrix for the system (8.2). Hence, there is a Stieltjes nonsingular matrix R such that
$$\Psi(t+\omega) = \Psi(t)R\omega, \quad t \in I.$$
Let
$$P(t) = \Psi(t)R^{-1}(t), \quad t \in I.$$
Then
$$P(t+\omega) = \Psi(t+\omega)R^{-1}(t+\omega)$$
$$= \Psi(t)R\omega R^{-1}(\omega)R^{-1}(t)$$
$$= \Psi(t)R^{-1}(t), \quad t \in I,$$

i.e.,
$$P(t+\omega) = P(t), \quad t \in I.$$
Thus, P is ω-periodic. Next, we have
$$\det \Psi(t) \neq 0, \quad t \in I,$$
and
$$\det R^{-1}(t) \neq 0, \quad t \in I.$$
Therefore,
$$\det P(\omega) = \det \Psi(t) \det R^{-1}(t)$$
$$\neq 0, \quad t \in I,$$
i.e., P is Stieltjes nonsingular. This completes the proof. □

Remark 8.1. Let Ψ be a fundamental matrix for the system (8.2) and P and R be as in Theorem 8.3. Then
$$\Psi(0) = P(0)$$
and
$$\Psi(\omega) = P(\omega)R\omega.$$
Hence, using the fact that
$$P(0) = P(\omega),$$
we get
$$\Psi(\omega) = P(0)R\omega$$
$$= \Psi(0)R\omega,$$
whereupon
$$R\omega = (\Psi(0))^{-1}\Psi(\omega),$$
This gives
$$R = \frac{(\Psi(0))^{-1}\Psi(\omega)}{\omega}.$$

Theorem 8.4. *Let Ψ be a fundamental matrix for the system (8.2), and P and R be as in Theorem 8.3. Then, the transformation*

$$x(t) = P(t)y(t), \quad t \in I,$$

reduces (8.2) to the system

$$y'_g = Ry \quad \text{on } I.$$

Proof. In view of Theorem 2.5, we have

$$x'_g(t) = P'_g(t)y(t) + P(t)y'_g(t), \quad t \in I. \tag{8.7}$$

Since x is a solution to the system (8.2), we have

$$x'_g(t) = A(t)y(t)$$
$$= A(t)P(t)y(t), \quad t \in I.$$

Hence, using (8.7), we get

$$A(t)P(t)y(t) = P'_g(t)y(t) + P(t)y'_g(t), \quad t \in I,$$

i.e.,

$$P(t)y'_g(t) = (A(t)P(t) - P'_g(t))y(t), \quad t \in I. \tag{8.8}$$

Since Ψ is a fundamental matrix for the system (8.2), we have

$$\Psi'_g(t) = A(t)\Psi(t), \quad t \in I,$$

and keeping in mind Theorem 8.3, we obtain

$$\Psi'_g(t) = P'_g(t)Rt + P(t)RRt$$
$$= (P'_g(t) + P(t)R)Rt, \quad t \in I.$$

Therefore,

$$(P'_g(t) + P(t)R)Rt = A(t)P(t)Rt, \quad t \in I,$$

i.e.,

$$P'_g(t) + P(t)R = A(t)P(t), \quad t \in I,$$

i.e.,
$$A(t)P(t) - P'_g(t) = P(t)R, \quad t \in I.$$
Hence, from (8.8), we find
$$P(x)y'_g(t) = P(t)Ry(t), \quad t \in I,$$
whereupon
$$y'_g(t) = Ry(t), \quad t \in I.$$
This completes the proof. □

Theorem 8.5. *Let Ψ be a fundamental matrix for the system (8.2) and P and R be as in Theorem 8.3. Then, for any fundamental matrix Ψ_1 of the system (8.2), there is a Stieltjes constant nonsingular matrix M such that*
$$\Psi_1(t+\omega) = \Psi_1(t)MR\omega M^{-1}, \quad t \in I. \tag{8.9}$$

Proof. By Theorem 8.3, we have
$$\Psi(t) = P(t)Rt,$$
$$\Psi(t+\omega) = P(t+\omega)R(t+\omega)$$
$$= P(t)RtR\omega$$
$$= \Psi(t)R\omega, \quad t \in I.$$
Since Ψ_1 is a fundamental matrix for the system (8.2), there is a Stieltjes constant nonsingular matrix M such that
$$\Psi(t) = \Psi_1(t)M, \quad t \in I, \tag{8.10}$$
whereupon
$$\Psi_1(t) = \Psi(t)M^{-1}, \quad t \in I.$$
Therefore,
$$\Psi_1(t+\omega) = \Psi(t+\omega)M^{-1}$$
$$= \Psi(t)R\omega M^{-1}$$
$$= \Psi_1(t)MR\omega M^{-1}, \quad t \in I.$$
This completes the proof. □

Theorem 8.6. *For any constant nonsingular matrix M, there is a fundamental matrix for the system (8.2) so that (8.9) holds.*

Proof. Let M be any constant nonsingular matrix. Then,

$$\Psi(t)M^{-1}, \quad t \in I,$$

is a fundamental matrix for (8.2). We set

$$\Psi_1(t) = \Psi(t)M^{-1}, \quad t \in I.$$

Then we get (8.10), and subsequently, (8.9). This completes the proof. □

Definition 8.1. Let Ψ be a fundamental matrix for the system (8.2), and P and R be as in Theorem 8.3. The matrix

$$R\omega$$

is said to be monodromy matrix for the system (8.2). The eigenvalues of matrix $R\omega$ are said to be Floquet multipliers of (8.2). The eigenvalues of matrix R are said to be exponents for the system (8.2).

Theorem 8.7. *Let Ψ be the Stieltjes transitive matrix for the system (8.2) and P and R be as in Theorem 8.3. Then, λ is a exponent for the system (8.2) if and only if there exists a nontrivial solution x of the system (8.2) of the form*

$$x(t) = \lambda t h(t), \quad t \in I, \tag{8.11}$$

where h is a ω-periodic function on I.

Proof. Suppose x is a nontrivial solution, for the system (8.2), of the form (8.11). Then,

$$x(t) = \Psi(t)x(t_0)$$
$$= \lambda t h(t)$$
$$= P(t)Rtx(t_0), \quad t \in I.$$

Hence,

$$x(t+\omega) = \lambda(t+\omega)h(t+\omega)$$
$$= \lambda(t+\omega)h(t)$$
$$= P(t+\omega)R(t+\omega)x(t_0)$$
$$= P(t)R(t+\omega)x(t_0)$$
$$= P(t)RtR\omega x(t_0), \quad t \in I,$$

i.e.,
$$\lambda(t+\omega)h(t) = P(t)RtR\omega x(t_0), \qquad (8.12)$$
$t \in I$. Note that
$$\lambda(t+\omega)h(t) = \lambda\omega\lambda t h(t)$$
$$= \lambda\omega x(t)$$
$$= \lambda\omega P(t)Rtx(t_0), \quad t \in I,$$
i.e.,
$$\lambda(t+\omega)h(t) = \lambda\omega P(t)Rtx(t_0), \quad t \in I.$$
Hence, from (8.12), we get
$$\lambda\omega P(t)Rtx(0) = P(t)RtR\omega x(t_0), \quad t \in I$$
i.e.,
$$P(t)Rt\left(\lambda\omega\mathscr{I} - R\omega\right)x(t_0) = 0, \quad t \in I,$$
whereupon
$$\det\left(\lambda\omega\mathscr{I} - R\omega\right) = 0, \quad t \in I,$$
i.e., λ is a exponent for the system (8.2). Conversely, suppose that λ is an exponent for the system (8.2). Take η to be a eigenvalue corresponding to the exponent λ, i.e.,
$$Rt\eta = \lambda t\eta, \quad t \in I.$$
Then
$$P(t)Rt\eta = P(t)\lambda t\eta, \quad t \in I.$$
Set
$$g(t) = P(t)\eta, \quad t \in \mathbb{R}.$$
Then,
$$P(t)\lambda t\eta = e_{\lambda,g}(t,t_0)h(t), \quad t \in I,$$
is a solution to the system (8.2). Note that
$$h(t+\omega) = P(t+\omega)\eta$$
$$= P(t)\eta$$
$$= h(t), \quad t \in \mathbb{R}.$$
This completes the proof. □

Exercise 8.1. Prove that the Stieltjes differential equation
$$x'''_g + x = \cos t, \quad t \in I,$$
has a ω-periodic solution.

Exercise 8.2. Prove that the Stieltjes differential system
$$x'_{1g} = 4x_1 - x_2 - x_3 - \sin t,$$
$$x'_{2g} = x_1 + 2x_2 - x_3 + \cos t,$$
$$x'_{3g} = x_1 - x_2 + 2x_3, \quad t \in I,$$
has no 2π-periodic solutions.

8.2 Asymptotic Behaviour of Solutions

Let $\mathbb{R}^+ = [0, \infty) = I$. In this section, we investigate the asymptotic behaviour of solutions of the Stieltjes differential system
$$x'_g = A(t)x + f(t), \quad t \in \mathbb{R}^+, \tag{8.13}$$
where $A \in \mathscr{C}(\mathbb{R}^+)$ is an $n \times n$-matrix with entries $a_{ij} \colon \mathbb{R}^+ \to \mathbb{R}$, $i, j \in \{1, \ldots, n\}$, and $f \in \mathscr{C}(\mathbb{R}^+)$ is an $n \times 1$-matrix with entries $f_j \colon \mathbb{R}^+ \to \mathbb{R}$, $j \in \{1, \ldots, n\}$.

Theorem 8.8. *Let all solutions of the system*
$$x'_g = A(t)x, \quad t \in \mathbb{R}^+, \tag{8.14}$$
are bounded and at least one solution of the system (8.13) *is bounded on* $\mathbb{R}^+$. *Then all solutions of the system* (8.13) *are bounded on* $\mathbb{R}^+$.

Proof. Let x_1 be a bounded solution and x_2 be any other solution of the system (8.13). Set
$$x_3(t) = x_1(t) - x_2(t), \quad t \in \mathbb{R}^+.$$
Then, x_3 is a solution of the system (8.14). Also, by hypothesis, x_3 is bounded on $\mathbb{R}^+$. Hence, x_2 is bounded on $\mathbb{R}^+$. This completes the proof. □

Exercise 8.3. Prove that all solutions of the Stieltjes system
$$x'_{1g} = -e^{-t}x_1 + \cos t,$$
$$x'_{2g} = e^{-3t}x_2 + t\cos t^2, \quad t \in \mathbb{R}^+,$$
are bounded.

8.3 Advanced Practical Problems

Problem 8.1. Prove that the function
$$f(t) = 3\sin\frac{t}{3} + 8\cos\frac{t}{6} + 2, \quad t \in I,$$
is 6π-periodic.

Problem 8.2. Prove that the following Stieltjes differential equation has no periodic solutions:

(1)
$$x_g'' + x = \left(\sin\frac{t}{4}\right)^4, \quad t \in I,$$

(2)
$$x_g'' - 2x_g' = 8(\sin t)^2, \quad t \in I,$$

(3)
$$x_g''' + x_g' = \sin t + (\sin t)^2, \quad t \in I,$$

(4)
$$x_g''' + 4x_g' = \sin t - 7(\sin t)^2, \quad t \in I,$$

(5)
$$x_g'' + x_g' = \sin t, \quad t \in I.$$

Problem 8.3. Prove that the following Stieltjes differential system has no 2π-periodic solutions:

(1)
$$x_{1g}' = x_1 - x_2 + x_3 - \cos t,$$
$$x_{2g}' = x_1 + x_2 - x_3 - \cos t,$$
$$x_{3g}' = 2x_1 - x_2, \quad t \in I.$$

(2)
$$x_{1g}' = x_1 - 2x_2 - x_3 + \sin t,$$
$$x_{2g}' = -x_1 + x_2 + x_3,$$
$$x_{3g}' = x_1 - x_3, \quad t \in I.$$

(3)
$$x'_{1g} = 2x_1 - x_2 + x_3 - 2\cos t,$$
$$x'_{2g} = x_1 + 2x_1 - x_3,$$
$$x'_{3g} = x_1 - x_2 + 2x_3 + \cos t, \quad x \in I.$$

(4)
$$x'_{1g} = 3x_1 - x_2 + x_3 - \sin t + \cos(2t),$$
$$x'_{2g} = x_1 + x_2 + x_3 + \sin t,$$
$$x'_{3g} = 4x_1 - x_2 + 4x_3, \quad t \in I.$$

(5)
$$x'_{1g} = x_1 - x_2 - x_3 - \cos t,$$
$$x'_{2g} = x_1 + x_2 - \cos t,$$
$$x'_{3g} = 3x_1 + x_3 - \cos t, \quad t \in I.$$

(6)
$$x'_{1g} = 2x_1 + x_2 + \sin(2t),$$
$$x'_{2g} = x_1 + 3x_2 - x_3 + \sin(4t),$$
$$x'_{3g} = -x_1 + 2x_2 + 3x_3 + \cos t, \quad t \in I.$$

(7)
$$x'_{1g} = 2x_1 - x_2 + 2x_3 - \sin t,$$
$$x'_{2g} = x_1 + 2x_3 - \sin t,$$
$$x'_{3g} = -2x_1 + x_2 - x_3 - \sin t, \quad t \in I.$$

(8)
$$x'_{1g} = -2x_1 + x_2 + 2x_3 - 3\cos t,$$
$$x'_{2g} = x_1 - 2x_2 + 2x_3 + \sin t,$$
$$x'_{3g} = 3x_1 - 3x_2 + 5x_3 - \cos t, \quad t \in I.$$

(9)
$$x'_{1g} = 3x_1 - 2x_2 - x_3 - \sin t + \cos t,$$
$$x'_{2g} = 3x_1 - 4x_2 - 3x_3 + \cos t,$$
$$x'_{3g} = 2x_1 - 4x_2 - \cos t, \quad t \in I.$$

(10)
$$x'_{1g} = x_1 - x_2 + x_3 + \cos t,$$
$$x'_{2g} = x_1 + x_2 - x_3 + \sin(8t),$$
$$x'_{3g} = -x_2 + 2x_3, \quad t \in I.$$

(11)
$$x'_{1g} = -x_1 + x_2 - 2x_3 - \cos t,$$
$$x'_{2g} = 4x_1 + x_2 - \sin t,$$
$$x'_{3g} = 2x_1 + x_2 - x_3, \quad t \in I.$$

(12)
$$x'_{1g} = -2x_1 + x_2 + \cos t,$$
$$x'_{2g} = 2x_2 + 4x_3 + \sin(4t),$$
$$x'_{3g} = x_1 - x_3, \quad t \in I.$$

(13)
$$x'_{1g} = 2x_1 - x_2 - x_3 - \cos t,$$
$$x'_{2g} = 2x_1 - x_2 - 2x_3,$$
$$x'_{3g} = -x_1 + x_2 + 2x_3, \quad t \in I.$$

(14)
$$x'_{1g} = 4x_1 - x_2 + \sin t + 3,$$
$$x'_{2g} = 3x_1 + x_2 - x_3,$$
$$x'_{3g} = x_1 + x_3 - \cos t, \quad t \in I.$$

(15)
$$x'_{1g} = 2x_1 -_8 x_2 - x_3 - \sin t,$$
$$x'_{2g} = 3x_1 - 2x_2 - 3x_3,$$
$$x'_{3g} = -x_1 + x_2 + 2x_3, \quad t \in I.$$

Problem 8.4. Prove that all solutions of the following Stieltjes differential systems are bounded:

(1)
$$x'_{1g} = x_1 - x_2,$$
$$x'_{2g} = 5x_1 - x_2, \quad t \in I.$$

(2)
$$x'_{1g} = x_2,$$
$$x'_{2g} = -x_1, \quad t \in I.$$

(3)
$$x'_{1g} = -4x_1 - 2x_2,$$
$$x'_{2g} = 6x_1 + 7x_2, \quad t \in I.$$

(4)
$$x'_{1g} = x_1 - 2x_2,$$
$$x'_{2g} = 2x_1 - 3x_2, \quad t \in I.$$

(5)
$$x'_{1g} = x_2 - x_3,$$
$$x'_{2g} = x_1 - x_3,$$
$$x'_{3g} = 2x_1 + 2x_2 - 3x_3, \quad t \in I.$$

Problem 8.5. Prove that the nontrivial solutions of the following Stieltjes differential systems are unbounded:

(1)
$$x''_{1g} + 3x''_{2g} - x_1 = 0,$$
$$x'_{1g} + 3x'_{2g} - 2x_2 = 0, \quad t \in I.$$

(2)
$$x''_{1g} + 5x'_{1g} + 2x'_{2g} + x_2 = 0,$$
$$3x''_{1g} + 5x_1 + x'_{2g} + 3x_2 = 0, \quad t \in I.$$

(3)
$$x''_{1g} + 4x'_{1g} - 2x_1 - 2x'_{2g} - x_2 = 0,$$
$$x''_{1g} - 4x'_{1g} - x''_{2g} + 2x'_{1g} + 2x_2 = 0, \quad t \in I.$$

(4)
$$2x''_{1g} + 2x'_{1g} + x_1 + 3x''_{2g} + x'_{2g} + x_2 = 0,$$
$$x''_{1g} + 4x'_{1g} - x_1 + 3x''_{2g} + 2x'_{2g} - x_2 = 0, \quad t \in I.$$

(5)
$$x'_{1g} = (2 + t)x_1 + x_2,$$
$$x'_{2g} = x_1 + 3x_2, \quad t \in I.$$

Problem 8.6. Prove that all solutions of the Stieltjes differential system

$$x'_{1g} = \cos t \; x_1 + 7x_2 + x_3,$$
$$x'_{2g} = -7x_1 + \cos t \; x_2 + 8x_3,$$
$$x'_{3g} = -x_1 - 8x_2 + \sin t \; x_3, \quad t \in I,$$

are bounded.

Problem 8.7. Prove that all solutions of the Stieltjes differential system

$$x'_{1g} = \cos t \; x_1 - \sin t \; x_2 - tx_4,$$
$$x'_{2g} = -\sin t \; x_1 - 3tx_2 - \cos(3t) \; x_5,$$
$$x'_{3g} = \sin(7t) \; x_5,$$
$$x'_{4g} = -2tx_4 - 4tx_2,$$
$$x'_{5g} = -tx_3 - 7tx_5, \quad t \in \mathbb{R}^+,$$

are bounded.

Problem 8.8. Prove that all solutions of the Stieltjes differential system

$$x'_{1g} = \sin t \; x_1 - \cos t \; x_6,$$
$$x'_{2g} = -2tx_3 - \cos t \; x_4 - \cos t \; x_6$$
$$x'_{3g} = -tx_2 - \sin(2t) \; x_5 - tx_6$$
$$x'_{4g} = -3tx_4 - tx_5 - \sin t \; x_6$$
$$x'_{5g} = -tx_1 - 4tx_5$$
$$x'_{6g} = -tx_1 - 3tx_6, \quad t \in \mathbb{R}^+,$$

are bounded.

Chapter 9

Stability Theory for Stieltjes Differential Systems

This chapter focuses on the stability analysis of Stieltjes differential systems. We give various criteria for the stability of solutions of Stieltjes differential systems and deduce sufficient conditions for uniform stability. We also investigate the stability of quasi-linear Stieltjes differential systems. Further, we present the stability of two-dimensional autonomous Stieltjes differential systems.

Suppose that $t_0 \in \mathbb{R}$ and $I = [t_0, \infty)$ and $g: I \to \mathbb{R}$ is monotone nondecreasing function that is continuous from the left everywhere.

9.1 Definition and Examples

We investigate the following Stieltjes initial value problem (SIVP):

$$x'_g = f(t, x), \quad t \in I,$$
$$x(t_0) = x_0, \tag{9.1}$$

where $t_0 \in \mathbb{R}$, $I = [t_0, \infty)$, and

$$x = \begin{pmatrix} x_1 \\ \vdots \\ x_n \end{pmatrix}, \quad f(t,x) = \begin{pmatrix} f_1(t,x) \\ \vdots \\ f_n(t,x) \end{pmatrix}, \quad \text{and} \quad x_0 = \begin{pmatrix} x_{01} \\ \vdots \\ x_{0n} \end{pmatrix}.$$

Definition 9.1. For an $m \times n$-matrix-valued function C with entries $c_{ij}: I \to \mathbb{R}$, $i \in \{1, \ldots, m\}$, $j \in \{1, \ldots, n\}$, define

$$\|C\| = \max_{\substack{i \in \{1,\ldots,m\} \\ j \in \{1,\ldots,n\}}} \left(\sup_{t \in I} |c_{ij}(t)| \right)$$

Definition 9.2. A solution $x(t) = x(t, t_0, x_0)$ of the SIVP (9.1) is said to be stable provided

(1) $x(t)$ exists on the interval I, and
(2) for every $\varepsilon > 0$, there is $\delta = \delta(\varepsilon, x_0) > 0$ such that $\|\Lambda x_0\| < \delta$ implies that

$$\|x(t, t_0, x_0 + \Lambda y_0) - x(t, t_0, x_0)\| < \epsilon$$

for all $t \in I$. Here, Λx_0 represents a small perturbation in x_0.

Otherwise, we state that $x(t) = x(t, t_0, x_0)$ is unstable.

Definition 9.3. A solution $x(t) = x(t, t_0, x_0)$ of the SIVP (9.1) is said to be asymptotically stable provided it is stable and there exists $\delta_0 > 0$ such that $\|\Lambda x_0\| < \delta_0$ implies that

$$\|x(t, t_0, x_0 + \Lambda x_0) - x(t, t_0, x_0)\| \to 0,$$

as $t \to \infty$.

Example 9.1. Consider the Stieltjes initial value problem

$$x'_g = a(t), \quad t \in I,$$
$$x(t_0) = x_0,$$

where $a \in \mathscr{C}(I)$. Its solution is given by

$$x(t) = x_0 + \int_{t_0}^{t} a(s) d_g s, \quad t \in I,$$

which is defined on I. Let $\varepsilon > 0$ be arbitrarily chosen and $\delta = \varepsilon$. Then for

$$|\Lambda x_0| < \delta = \varepsilon,$$

we have

$$|x(t,t_0,x_0+\Lambda x_0) - x(t,t_0,x_0)| = \left|x_0 + \lambda t_0 + \int_{t_0}^t a(s)d_g s - x_0 \right.$$
$$\left. - \int_{t_0}^t a(s)d_g s\right|$$
$$= |\Lambda y_0|$$
$$< \varepsilon$$

for $t \in I$. Consequently, x is stable.

Exercise 9.1. Using the definition for stable/unstable solution, investigate for stability the solution of the following SIVPs:

(1)
$$x'_g = 4x - t^2 x, \quad t \in [1, \infty),$$
$$x(1) = 1.$$

(2)
$$x'_g = t - x, \quad t \in [1, \infty),$$
$$x(1) = e.$$

(3)
$$2tx'_g = x - x^3, \quad t \in [e, \infty),$$
$$x(e) = 1.$$

9.2 Criteria for Stability

Theorem 9.1. *All solutions of the Stieltjes differential system*

$$x'_g = A(t)x, \quad t \in I, \tag{9.2}$$

where A is an $n \times n$-matrix-valued continuous function on I, are stable if and only if they are bounded in I.

Proof. Let Ψ be a fundamental matrix of (9.2). Then every solution of (9.2) for which $x(t_0) = x_0$ can be represented in the form

$$x(t) = x(t, t_0, x_0)$$
$$= \Psi(t)\Psi^{-1}(t_0)x_0, \quad t \in I,$$

and it is defined on I. Suppose all solutions of the system (9.2) be stable. In particular, the trivial solution $y(x, x_0, y_0) = 1$ is stable. Then for every $\varepsilon > 0$, there exists $\delta = \delta(\varepsilon) > 0$ such that

$$\|\Lambda x_0\| < \delta$$

implies that

$$\|x(t, t_0, \Lambda x_0)\| < \varepsilon \quad \text{for all } t \geq t_0.$$

However,

$$x(t, t_0, \Lambda x_0) = \Psi(t)\Psi^{-1}(t_0)\Lambda x_0, \quad t \geq t_0.$$

Therefore,

$$\|\Psi(t)\Psi^{-1}(t_0)\Lambda x_0\| < \varepsilon, \quad t \geq t_0.$$

Let $\Lambda x_0 = (\delta/2)j$. Then

$$\|\Psi(t)\Psi^{-1}(t_0)\Lambda x_0\| = \|\Psi^j(t)\|\frac{\delta}{2}$$
$$< \varepsilon, \quad t \geq t_0,$$

where $\Psi^j(t)$ is the jth column of $\Psi(t)(\Psi(t_0))^{-1}$. This gives

$$\|\Psi^j(t)\|\frac{\delta}{2} < \varepsilon, \quad t \geq t_0.$$

Consequently,

$$\max_{1 \leq j \leq n} \|\Psi^j(t)\| < \frac{2\varepsilon}{\delta}, \quad t \geq t_0,$$

and hence, for any solution $x(t, t_0, x_0)$, we have that

$$\|x(t, t_0, x_0)\| = \|\Psi(t)\Psi^{-1}(t_0)x_0\|$$
$$< \frac{2\varepsilon}{\delta}\|x_0\|, \quad t \geq t_0,$$

Thus, all solutions of (9.2) are bounded.

Conversely, suppose that all solutions of (9.2) are bounded. Then, there exists a positive constant c such that
$$||\Psi(t)|| \le c, \quad t \in I.$$
Let $\varepsilon > 0$ be arbitrarily chosen and
$$\delta = \delta(\varepsilon, t_0)$$
$$= \frac{\varepsilon}{(c||(\Psi(t_0))^{-1}||)}.$$
Hence, for $t \in I$, the inequality
$$||\Lambda x_0|| < \delta$$
implies
$$||x(t, t_0, x_0 + \Lambda x_0) - x(t, t_0, x_0)|| = ||\Psi(t)\Psi^{-1}(t_0)\Lambda x_0||$$
$$\le ||\Psi(t)||\,||\Psi^{-1}(t_0)||\,||\Lambda x_0||$$
$$\le c||\Psi^{-1}(t_0)||\,||\Lambda x_0||$$
$$< \varepsilon,$$
i.e.,
$$||x(t, t_0, x_0 + \Lambda x_0) - x(t, t_0, x_0)|| < \varepsilon, t \ge t_0.$$
Thus, all solutions of (9.2) are stable. This completes the proof. □

We immediately have the following corollary.

Corollary 9.1. *If the real parts of the multiple eigenvalues of the matrix A are negative, and the real parts of the simple eigenvalues of the matrix A are nonpositive, then all solutions of the differential system*
$$x'_g = Ax \tag{9.3}$$
are stable.

Example 9.2. Consider the Stieltjes differential system
$$x'_{1g} = -x_1 + 5x_2,$$
$$x'_{2g} = -x_1 - x_2.$$

Here
$$A = \begin{pmatrix} -1 & 5 \\ -1 & -1 \end{pmatrix}.$$

Then
$$\det(A - \lambda \mathscr{I}) = \det \begin{pmatrix} -1-\lambda & 5 \\ -1 & -1-\lambda \end{pmatrix}$$
$$= (\lambda+1)^2 + 5$$
$$= \lambda^2 + 2\lambda + 6,$$

whereupon $\det(A - \lambda \mathscr{I}) = 0$ yields $\lambda_{1,2} = -1 \pm i\sqrt{5}$. Since $\operatorname{Re}(\lambda_{1,2}) < 0$, by Corollary 9.1, we state that all solutions of the considered Stieltjes differential system are stable.

Example 9.3. Consider the Stieltjes differential system
$$x'_{1g} = x_2,$$
$$x'_{2g} = x_3,$$
$$x'_{3g} = -3x_2 - 2x_3, \quad t \in I.$$

Here
$$A = \begin{pmatrix} 0 & 1 & 0 \\ 0 & 0 & 1 \\ 0 & -3 & -2 \end{pmatrix}.$$

Then
$$\det(A - \lambda \mathscr{I}) = \det \begin{pmatrix} -\lambda & 1 & 1 \\ 0 & -\lambda & 1 \\ 0 & -3 & -2-\lambda \end{pmatrix}$$
$$= -\lambda^2(\lambda - 2) - 3\lambda$$
$$= -\lambda(\lambda^2 + 2\lambda + 3),$$

whereupon $\det(A - \lambda \mathscr{I}) = 0$ yields $\lambda_1 = 0$ and $\lambda_{2,3} = -1 \pm 2i$. Consequently, in view of Corollary 9.1, all solutions of the considered system are stable.

Example 9.4. Consider the Stieltjes differential system

$$x'_{1g} = x_2,$$
$$x'_{2g} = x_3$$
$$x'_{3g} = -x_2 - 2x_3, \quad t \in I.$$

Here

$$A = \begin{pmatrix} 0 & 1 & 0 \\ 0 & 0 & 1 \\ 0 & -1 & -2 \end{pmatrix}.$$

Then

$$\det(A - \lambda \mathscr{I}) = \det \begin{pmatrix} -\lambda & 1 & 0 \\ 0 & -\lambda & 1 \\ 0 & -1 & -2-\lambda \end{pmatrix}$$
$$= -\lambda^2(\lambda + 2) - \lambda$$
$$= -\lambda(\lambda^2 + 2\lambda + 1)$$
$$= -\lambda(\lambda + 1)^2,$$

whereupon $\det(A - \lambda\mathscr{I}) = 0$ yields $\lambda_1 = 0$ and $\lambda_{2,3} = -1$. Consequently, in view of Corollary 9.1, all solutions of the considered system are stable.

Exercise 9.2. Prove that all solutions of the Stieltjes differential system

$$x'_{1g} = x_2,$$
$$x'_{2g} = x_3,$$
$$x'_{3g} = -6x_1 - 11x_2 - 6x_3, \quad t \in I,$$

are stable.

Theorem 9.2. *Let Ψ be a fundamental matrix of the system (9.2). Then, all solutions of the system (9.2) are asymptotically stable if and only if*

$$\|\Psi(t)\| \to 0 \quad \text{as } t \to \infty. \tag{9.4}$$

Proof. Suppose that (9.4) holds. Since the fundamental matrix Ψ is continuous, there exists constant $c > 0$ such that

$$\|\Psi(t)\| \leq c, \quad t \in I.$$

Every solution of the system (9.2) for which $x(t_0) = x_0$ is represented in the form

$$x(t, t_0, x_0) = \Psi(t)(\Psi(t_0))^{-1}x_0, \quad t \geq t_0.$$

Hence,

$$\begin{aligned}\|x(t, t_0, x_0)\| &= \|\Psi(t)(\Psi(t_0))^{-1}x_0\| \\ &\leq \|\Psi(t)\|\|(\Psi(t_0))^{-1}x_0\| \\ &\leq c\|(\Psi(t_0))^{-1}x_0\|\end{aligned}$$

for all $t \in I$, i.e., all solutions of the system (9.2) are bounded in I. In view of Theorem 9.1, we conclude that all solutions of (9.2) are stable. Also,

$$\begin{aligned}\|x(t, t_0, x_0 + \Lambda x_0) - x(t, t_0, x_0)\| &= \|\Psi(t)(\Psi(t_0))^{-1}\Lambda x_0\| \\ &\leq \|\Psi(t)\|\|(\Psi(t_0))^{-1}\Lambda x_0\| \\ &\to 0 \quad \text{as} \quad t \to \infty.\end{aligned}$$

Thus, all solutions of the system (9.2) are asymptotically stable. Conversely, suppose that all solutions of the system (9.2) are asymptotically stable. Then, in particular, the zero solution is asymptotically stable. Hence

$$\|x(t, t_0, \Lambda x_0)\| = \|\Psi(t)(\Psi(t_0))^{-1}\Lambda x_0\|$$
$$\to 0 \quad \text{as } t \to \infty,$$

whereupon (9.4) holds. This completes the proof. $\square$

We immediately have the following corollary.

Corollary 9.2. *If the real parts of the eigenvalues of the matrix A are negative, then all solutions of the system (9.3) are asymptotically stable.*

Example 9.5. Consider the Stieltjes differential system

$$x'_{1g} = x_2,$$
$$x'_{2g} = x_3,$$
$$x'_{3g} = -x_1 - 2x_2 - 2x_3, \quad t \in I.$$

Here

$$A = \begin{pmatrix} 0 & 1 & 0 \\ 0 & 0 & 1 \\ -1 & -2 & -2 \end{pmatrix}.$$

Then

$$\det(A - \lambda \mathscr{I}) = \det \begin{pmatrix} -\lambda & 1 & 0 \\ 0 & -\lambda & 1 \\ -1 & -2 & -2-\lambda \end{pmatrix}$$
$$= -\lambda^2(\lambda + 2) - 1 - 2\lambda$$
$$= -\lambda(\lambda(\lambda + 2) + 1) - (1 + \lambda)$$
$$= -\lambda(\lambda^2 + 2\lambda + 1) - (1 + \lambda)$$
$$= -\lambda(\lambda + 1)^2 - (\lambda + 1)$$
$$= -(\lambda + 1)(\lambda^2 + \lambda + 1),$$

whereupon $\det(A - \lambda \mathscr{I}) = 0$ yields $\lambda_1 = -1$ and $\lambda_{2,3} = (-1/2) \pm (i\sqrt{3}/2)$. Consequently, in view of Corollary 9.2, all solutions are asymptotically stable.

Example 9.6. Consider the Stieltjes differential system

$$x'_{1g} = -2x_1 - 7x_2,$$
$$x'_{2g} = x_1 - 4x_2, \quad t \in I.$$

Here

$$A = \begin{pmatrix} -2 & -7 \\ 1 & -4 \end{pmatrix}.$$

Then

$$\det(A - \lambda \mathscr{I}) = \det \begin{pmatrix} -2 - \lambda & -7 \\ 1 & -4 - \lambda \end{pmatrix}$$
$$= (\lambda + 4)(\lambda + 2) + 7$$
$$= \lambda^2 + 6\lambda + 15,$$

whereupon $\det(A - \lambda \mathscr{I}) = 0$ yields $\lambda_{1,2} = -\pm i\sqrt{6}$. Consequently, in view of Corollary 9.2, all solutions are asymptotically stable.

Exercise 9.3. Prove that all solutions of the Stieltjes differential system

$$x'_{1g} = -x_1 + 10x_2,$$
$$x'_{2g} = -3x_1 - 4x_2, \quad t \in I,$$

are asymptotically stable.

Theorem 9.3. *Let all solutions of the system (9.3) be stable and*

$$\int_{t_0}^{\infty} \|B(s)\| d_g s < \infty. \tag{9.5}$$

Then all solutions of the system

$$x'_g = (A + B(t))x, \quad t \in I, \tag{9.6}$$

are stable. Here, B is an $n \times n$-matrix-valued continuous function on I.

Proof. Since the condition (9.5) holds, by our hypothesis, we have that all solutions of the system (9.6) are bounded. Hence, in view of Theorem 9.1, all solutions of the system (9.6) are stable. This completes the proof. □

Theorem 9.4. *Let all solutions to the system (9.3) are asymptotically stable and*

$$\|B(t)\| \to 0 \quad \text{as } t \to \infty, \tag{9.7}$$

where B is an $n \times n$-matrix-valued continuous function on I. Then all solutions of the system (9.6) are asymptotically stable.

Proof. Since all solutions of the system (9.3) are asymptotically stable, in view of Definition 9.3, we have that all solutions of the system (9.3) tend to zero as $t \to \infty$. Hence, from the condition (9.7), we conclude that all solutions of the system (9.6) tend to zero as $t \to \infty$. If Ψ is a fundamental matrix of (9.6), then

$$||\Psi(t)|| \to 0$$

as $t \to \infty$. Consequently, in view of Theorem 9.2, all solutions to the system (9.6) are asymptotically stable. □

9.3 Uniform Stability

Definition 9.4. A solution $x(t) = x(t, t_0, x_0)$ of the SIVP (9.1) is said to be uniformly stable provided for each $\varepsilon > 0$, there is $\delta = \delta(\varepsilon) > 0$ such that, for any solution $x_*(t, t_0, x_1)$ of the SIVP

$$x'_g = f(t, x), \quad t \in I,$$
$$x(t_0) = x_1,$$

the inequalities $t_1 \geq t_0$ and

$$||x_*(t_1) - x(t_1)|| < \delta$$

imply that

$$||x_*(t) - x(t)|| < \varepsilon$$

for all $t \geq t_1$.

Theorem 9.5. *Let Ψ be a fundamental matrix of the system (9.2). Then, all solutions of the (9.2) are uniformly stable if and only if*

$$||\Psi(t)(\Psi(s))^{-1}|| \leq c, \quad t_0 \leq s \leq t < \infty, \qquad (9.8)$$

for some positive constant c.

Proof. Suppose that the condition (9.8) holds. Let $x(t) = x(t, t_0, x_0)$ be a solution to the system (9.2) Then, for any $t_1 \geq t_0$, we have

$$x(t) = \Psi(t)(\Psi(t_1))^{-1} x(t_1).$$

Assume that
$$x_1(t) = \Psi(t)(\Psi(t_1))^{-1}x_1(t_1), \quad t_1 \geq t_0$$
is any other solution of the system (9.2). Then, for $t \geq t_1$, we have
$$\begin{aligned}||x_1(t) - x(t)|| &= ||\Psi(t)(\Psi(t_1))^{-1}x(t_1) - \Psi(t)(\Psi(t_1))^{-1}x_1(t_1)|| \\ &= ||\Psi(t)(\Psi(t_1))^{-1}(x(t_1) - x_1(t_1))|| \\ &\leq ||\Psi(t)(\Psi(t_1))^{-1}||||x(t_1) - x_1(t_1)|| \\ &\leq c||x(t_1) - x_1(t_1)||,\end{aligned}$$
i.e.,
$$||x_1(t) - x(t)|| \leq c||x(t_1) - x_1(t_1)||, \quad \text{for } t \geq t_1.$$
Let $\varepsilon > 0$ be arbitrarily chosen. Then the inequalities $t_1 \geq t_0$ and
$$||x(t_1) - x_1(t_1)|| < \frac{\varepsilon}{c}$$
imply
$$||x_1(t) - x(t)|| < \varepsilon \quad \text{for all } t \geq t_1.$$
Hence, the solution x is uniformly stable.

Conversely, suppose all solutions of the system (9.2) are uniformly stable. In particular, the zero solution $x(t, t_0, 0) = 0$ is uniformly stable. Therefore, for given $\varepsilon > 0$, there is $\delta = \delta(\varepsilon) > 0$ such that the inequalities $t_1 \geq t_0$ and
$$||x_1(t_1)|| < \delta$$
imply
$$||x_1(t)|| < \epsilon$$
for any $t \geq t_1$, i.e.,
$$||\Psi(t)\Psi^{-1}(t_1)x_1(t_1)|| < \varepsilon \quad \text{for all } t \geq t_1.$$
Let $x_1(t_1) = (\delta/2)j$. Then, from the last inequality, we get
$$||\Psi(t)(\Psi(t_1))^{-1}x_1(t_1)|| = ||\psi^j(t)||(\delta/2)$$
$$< \varepsilon, \quad t \geq t_1,$$

where ψ^j is the jth column of $\Psi(t)(\Psi(t_1))^{-1}$. From here,

$$||\Psi(t)(\Psi(t_1))^{-1}|| = \max_{1\leq j\leq n} ||\psi^j(t)||$$
$$< (2\varepsilon)/\delta, \quad t \geq t_1.$$

This completes the proof. □

Now, we recall the concept of Hurwitz matrix and the Hurwitz criteria. For a given polynomial

$$p(t) = a_0 t^n + a_1 t^{n-1} + \cdots + a_{n-1} t + a_n, \quad t \in I, \quad a_i \in \mathbb{C},$$
$$i \in \{0, 1, \ldots, n\}, \tag{9.9}$$

the $n \times n$ square matrix

$$H = \begin{pmatrix} a_1 & a_3 & a_5 & \cdots & 0 & 0 & 0 \\ a_0 & a_2 & a_4 & \cdots & & & \\ 0 & a_1 & a_3 & \cdots & & & \\ \vdots & & & & & & \\ 0 & 0 & 0 & \cdots & a_{n-3} & a_{n-1} & 0 \end{pmatrix}$$

is said to be the Hurwitz matrix of (9.9).

Theorem 9.6 (The Hurwitz Criteria). *If $a_0 > 0$, then the polynomial (9.9) is stable (i.e., all its roots have strictly negative real part) if and only if all the leading principal minors of the matrix H are positive.*

Exercise 9.4. Use the Hurwitz criteria to find the parameter a in the Stieltjes differential system

$$x'_{1g} = x_3,$$
$$x'_{2g} = -3x_1,$$
$$x'_{3g} = ax_1 + 2x_2 - x_3, \quad t \in I,$$

so that the trivial solution is asymptotically stable.

Solution. Here

$$A = \begin{pmatrix} 0 & 0 & 1 \\ -3 & 0 & 0 \\ a & 2 & -1 \end{pmatrix}.$$

Then

$$\det(A - \lambda \mathscr{I}) = \det \begin{pmatrix} -\lambda & 0 & 1 \\ -3 & -\lambda & 0 \\ a & 2 & -1-\lambda \end{pmatrix}$$

$$= -\lambda^2(\lambda+1) - 6 + a\lambda$$

$$= -\lambda^3 - \lambda^2 + a\lambda - 6.$$

We consider the polynomial

$$p(\lambda) = \lambda^3 + \lambda^2 - a\lambda + 6.$$

The corresponding Hurwitz matrix is

$$\begin{pmatrix} 1 & 1 & 0 \\ 6 & -a & 1 \\ 0 & 0 & 6 \end{pmatrix}.$$

Here $a_0 = 1 > 0$ and

$$\det \begin{pmatrix} 1 & 1 \\ 6 & -a \end{pmatrix} = -a - 6$$

$$> 0 \quad \text{provided} \quad a < -6.$$

Now, since

$$\det \begin{pmatrix} 1 & 1 & 0 \\ 6 & -a & 1 \\ 0 & 0 & 6 \end{pmatrix} = 6 \det \begin{pmatrix} 1 & 1 \\ 6 & -a \end{pmatrix}$$

$$> 0 \quad \text{provided} \quad a < -6,$$

in view of Hurwitz criteria, we can say that the trivial solution is asymptotically stable provided $a < -6$. □

Exercise 9.5. Use the Hurwitz criteria to find the real parameters a and b such that the trivial solution of the system

$$x'_{1g} = x_1 + x_2 - x_3,$$
$$x'_{2g} = ax_1 + x_2 + x_3,$$
$$x'_{3g} = -x_1 + bx_2 - 3x_3, \quad t \in I,$$

is asymptotically stable.

Answer 9.1. The required real parameters a and b are such that

$$2a - b - ab - 11 > 0$$

and

$$-3a + b + ab + 5 > 0.$$

Theorem 9.7. *Let the $n \times n$-matrix A and the $n \times 1$-vector b be continuous on I. If all solutions of the system*

$$x'_g = A(t)x + b(t) \tag{9.10}$$

are bounded on I, then they are stable.

Proof. Let Ψ be a fundamental matrix of the system (9.2). Then the solution x of the system (9.10) for which $x(t_0) = x_0$ can be represented in the form

$$y(x) = \Psi(x)(\Psi(x_0))^{-1} y_0 + \int_{x_0}^{x} \Psi(x)(\Psi(t))^{-1} b(t) d_g t.$$

Since all solutions of the system (9.10) are bounded, there is a constant $c > 0$ such that

$$\|\Psi(t)\| \le c, \quad t \in I.$$

Let $\varepsilon > 0$ be arbitrarily chosen and

$$\delta = \delta(\varepsilon, t_0) = \frac{\varepsilon}{(c\|(\Psi(t_0))^{-1}\|)}.$$

Then, for

$$\|\Lambda x_0\| < \delta,$$

we have
$$\|x(t,t_0,x_0+\Lambda x_0)-x(t,t_0,x_0)\| = \|\Psi(t)(\Psi(t_0))^{-1}\Lambda x_0\|$$
$$\leq \|\Psi(t)\|\|(\Psi(t_0))^{-1}\|\|\Lambda x_0\|$$
$$\leq c\|(\Psi(t_0))^{-1}\|\|\Lambda x_0\|$$
$$< \varepsilon.$$

i.e.,
$$\|x(t,t_0,x_0+\Lambda x_0)-x(t,t_0,x_0)\| < \varepsilon, \quad t \geq t_0.$$

Thus, all solutions of (9.10) are stable. This completes the proof. □

Theorem 9.8. *Let the $n \times n$-matrix A and the $n \times 1$-vector b be continuous on I. If all solutions of the system (9.10) are stable and one solution is bounded, then all solutions of the system (9.10) are bounded on I.*

Proof. Let Ψ be a fundamental matrix of the system (9.2). Let $x(t) = x(t,t_0,x_0)$ be the solution of the system (9.10) for which $x(t_0) = x_0$. Also, let $\varepsilon > 0$ be arbitrarily chosen. Then there is $\delta = \delta(\varepsilon,t_0) > 0$ such that the inequality
$$\|\Lambda x_0\| < \delta$$
implies
$$\|x(t,t_0,x_0+\Lambda x_0)-x(t,t_0,x_0)\| = \|\Psi(t)\Psi^{-1}(t_0)\Delta x_0\|$$
$$< \varepsilon, \quad t \in I. \tag{9.11}$$

We take $\Lambda x_0 = (\delta/2)j$. Then, from (9.11), we get
$$\|x(t,t_0,x_0+\Lambda x_0)-x(t,t_0,x_0)\| = \|\Psi^j(t)\|\frac{\delta}{2}$$
$$< \varepsilon, \quad t \geq t_0,$$

where $\Psi^j(t)$ is the jth column of $\Psi(t)(\Psi(t_0))^{-1}$. Consequently,
$$\max_{1 \leq j \leq n} \|\Psi^j(t)\| < \frac{2\varepsilon}{\delta}, \quad t \in I.$$

Hence, all solutions of the system (9.2) are bounded on I. From here, using the fact that one solution of (9.10) is bounded, we conclude that all solutions of (9.10) are bounded. This completes the proof. □

Theorem 9.9. *Let the $n \times n$-matrix A and the $n \times 1$-vector b be continuous on I. Then the stability of any solution of the nonhomogeneous system (9.10) is equivalent to the stability of the trivial solution of the system (9.2).*

Proof. Let Ψ be a fundamental matrix of the system (9.2). Using Definition 9.2, any solution $x(t, t_0, x_0)$ of the system (9.10) for which $x(t_0) = x_0$ is stable if and only if for every $\varepsilon > 0$, there is $\delta = \delta(\varepsilon, t_0) > 0$ such that the inequality

$$\|\Lambda x_0\| < \delta$$

implies

$$\|\Psi(t)(\Psi(t_0))^{-1} x_0\| < \varepsilon, \quad t \in I,$$

if and only if the trivial solution of (9.2) is stable. This completes the proof. □

Theorem 9.10. *Let A be an $n \times n$-matrix-valued continuous, ω-periodic function on $\mathbb{R}$. If the Floquet multipliers σ_i of the system (9.2) satisfy $|\sigma_i| < 1$, $i \in \{1, 2, \ldots, n\}$, then the trivial solution of (9.2) is asymptotically stable.*

Proof. Let λ_i, $i \in \{1, 2, \ldots, n\}$, be the exponents of the system (9.2). Since

$$\sigma_i = \lambda_i t \quad \text{and} \quad |\sigma_i| < 1, \quad i \in \{1, 2, \ldots, n\},$$

we conclude that

$$\operatorname{Re}(\lambda_i) < 0.$$

Hence, in view of Corollary 9.2, the trivial solution of the system (9.2) is asymptotically stable. This completes the proof. □

Definition 9.5. *Let the $n \times n$-matrix A be continuous on I. The system (9.2) is said to be stable provided all its solutions are stable.*

Definition 9.6. Let the $n \times n$-matrix A be continuous on I. The system (9.2) is called restrictively stable provided the system (9.2) together with its adjoint system

$$x'_g = -(A(t))^T x, \quad t \in I, \tag{9.12}$$

are stable.

Theorem 9.11. *Let the $n \times n$-matrix A be continuous on I. Then a necessary and sufficient condition for restrictive stability of (9.2) is that there exists constant $c > 0$ such that*

$$\|\Phi(t, t_0)\Phi(t_0, s)\| \leq c, \quad t \geq t_0, \quad s \geq t_0, \tag{9.13}$$

where $\Phi(t, t_0)$ is the transitive matrix of the system (9.2).

Proof. Suppose the system (9.2) be restrictively stable. Then the systems (9.2) and (9.12) are stable. Hence, all solutions of the systems (9.2) and (9.12) are bounded. Since $\Phi(t, t_0)$ is the transitive matrix of (9.2), we have that

$$\left(\Phi(s, t_0)^{-1}\right)^T = \Phi(t_0, t)^T, \quad t, s \geq t_0,$$

is the Stieltjes transitive matrix of the system (9.12). Since all solutions of (9.2) and (9.12) are bounded, there is a constant $c > 0$ such that

$$\|\Phi(t, t_0)\| \leq c^{\frac{1}{2}},$$

and

$$\|\Phi(s, t_0)^T\| = \|\Phi(s, t_0)\|$$
$$\leq c^{\frac{1}{2}}, \quad s, t \geq t_0.$$

Hence,

$$\|\Phi(t, t_0)\Phi(s, t_0)\| \leq \|\Phi(t, t_0)\|\|\Phi(s, t_0)\|$$
$$\leq c^{\frac{1}{2}} c^{\frac{1}{2}}$$
$$= c, \quad s, t \geq t_0.$$

Conversely, suppose (9.13) holds. Then, the Stieltjes transitive matrices $\Phi(t, t_0)$ and $\Phi(s, t_0)^T$ of the systems (9.2) and (9.12), respectively, are bounded on I. Hence, all solutions of (9.2) and (9.12) are bounded on I. Therefore, in view of Theorem 9.1, all solutions of the systems (9.2) and (9.12) are stable. Hence, the system (9.2) is restrictively stable. This completes the proof. □

Theorem 9.12. *Let the $n \times n$-matrix A be continuous on I. Also, let the system (9.2) be stable and*

$$\liminf_{t \to \infty} \int_{t_0}^{t} \operatorname{tr} A(s) ds > -\infty \quad or \quad \operatorname{tr} A(t) = 0, \quad t \geq t_0. \tag{9.14}$$

Then the system (9.2) is restrictively stable.

Proof. Let Ψ be a fundamental matrix of the system (9.2). Using the Stieltjes–Abel Theorem II, Theorem 7.15, we have that

$$\det \Psi(t) = \det \Psi(t_0) \int_{t_0}^{t} \operatorname{tr} A(s) d_g s, \quad t \geq t_0,$$

and hence,

$$(\Psi(t))^{-1} = \frac{\operatorname{adj} \Psi(t)}{\det \Psi(t)}$$

$$= \frac{\operatorname{adj} \Psi(t)}{\det \Psi(t_0) \int_{t_0}^{t} \operatorname{tr} A(s) d_g s}, \quad t \geq t_0.$$

Based on our assumption (9.14), we conclude that $\|(\Psi(t))^{-1}\|$ is bounded on I, whereupon $\|((\Psi(t))^{-1})^T\|$ is bounded on I. Since $((\Psi(t))^{-1})^T$ is a fundamental matrix of the system (9.12), we conclude that all solutions of the system (9.12) are bounded on I. Therefore, in view of Theorem 9.1, all solutions of (9.12) are stable, i.e., the system (9.12) is stable. Consequently, the system (9.2) is restrictively stable. This completes the proof. □

Theorem 9.13. *Let the $n \times n$-matrix A be continuous on I. Suppose the system (9.12) is stable and*

$$\limsup_{t \to \infty} \int_{t_0}^{t} \operatorname{tr} A(s) d_g s < \infty. \tag{9.15}$$

Then the system (9.2) is restrictively stable.

Proof. Let Ψ be a fundamental matrix of (9.2). Then, using the Stieltjes–Abel Theorem II, Theorem 7.15, we have that

$$\det\Psi(t) = \det\Psi(t_0)\int_{t_0}^{t} \mathrm{tr}A(s)d_g s, \quad t \geq t_0.$$

Hence, from (9.15), $\|\Psi(t)\|$ is bounded on I. Thus, all solutions of (9.2) are bounded on I. Therefore, in view of Theorem 9.1, the system (9.2) is stable. Consequently, the system (9.2) is restrictively stable. This completes the proof. $\square$

9.4 Stability of Quasi-Linear Stieltjes Differential Systems

In this section, we investigate the following Stieltjes differential systems:

$$x'_g = Ax + f(t, x), \qquad (9.16)$$

$$x'_g = B(t)x + f(t, x), \quad t \in I, \qquad (9.17)$$

where $f\colon I \times \mathbb{R}^n \to \mathbb{R}^n$ is continuous function, A is an $n \times n$ constant matrix, and B is an $n \times n$ continuous matrix on I.

Definition 9.7. The systems (9.16) and (9.17) will be called quasi-linear Stieltjes systems.

We suppose that the function $f\colon I \times \mathbb{R}^n \to \mathbb{R}^n$ satisfies the condition

$$\|f(t, x)\| = o(\|x\|) \qquad (9.18)$$

uniformly in t as $\|x\|$ approaches 0. Then, for every $\varepsilon > 0$, there exists $\delta = \delta(\varepsilon) > 0$ such that the inequality $\|x\| < \delta$ implies the inequality

$$\frac{\|f(t, x)\|}{\|x\|} < \varepsilon.$$

Hence, $f(t, 0) = 0$. Consequently, $x = 0$ is a solution to the quasi-linear Stieltjes differential systems (9.16) and (9.17).

Example 9.7. Consider the quasi-linear Stieltjes differential system

$$x'_{1g} = -x_1 - 5x_2 + x_1^2 + x_2^2,$$
$$x'_{2g} = x_1 - 2x_2, \quad t \in I.$$

Here,

$$A = \begin{pmatrix} -1 & -5 \\ 1 & -2 \end{pmatrix}, \quad f(t,x) = \begin{pmatrix} x_1^2 + x_2^2 \\ 1 \end{pmatrix}, \quad t \in I.$$

Then, $f(t,x) = o(||x||)$ uniformly in t as $||x|| \to 0$. Next,

$$\det(A - \lambda \mathscr{I}) = \det \begin{pmatrix} -1-\lambda & -5 \\ 1 & -2-\lambda \end{pmatrix}$$
$$= (\lambda+2)(\lambda+1) + 5$$
$$= \lambda^2 + 3\lambda + 7,$$

whereupon $\det(A - \lambda \mathscr{I}) = 0$ yields

$$\lambda_{1,2} = \frac{-3 \pm i\sqrt{19}}{2}.$$

We find that $\operatorname{Re}(\lambda_{1,2}) < 0$. Hence, in view of Corollary 9.2, the trivial solution of the considered system is asymptotically stable.

Example 9.8. Consider the quasi-linear Stieltjes differential system

$$x'_{1g} = x_2 + x_1^4,$$
$$x'_{2g} = x_3 + x_1^4 + 2x_1^2 x_2^2 x_3^2,$$
$$x'_{3g} = -3x_1 - 5x_2 - 3x_3 + x_3^4, \quad t \in I.$$

Here,

$$A = \begin{pmatrix} 0 & 1 & 0 \\ 0 & 0 & 1 \\ -3 & -5 & -3 \end{pmatrix}, \quad f(t,x) = \begin{pmatrix} x_1^4 \\ x_1^4 + 2x_1^2 x_2^2 x_3^2 \\ x_3^4 \end{pmatrix}, \quad t \in I.$$

Then, $f(t,x) = o(||x||)$ uniformly in t as $||x|| \to 0$. Next,

$$\det(A - \lambda \mathscr{I}) = \det \begin{pmatrix} -\lambda & 1 & 0 \\ 0 & -\lambda & 1 \\ -3 & -5 & -3-\lambda \end{pmatrix}$$

$$= -\lambda^2(\lambda + 3) - 3 - 5\lambda$$
$$= -(\lambda^3 + 3\lambda^2 + 5\lambda + 3)$$
$$= -(\lambda^3 + \lambda^2 + 2\lambda^2 + 2\lambda + 3\lambda + 3)$$
$$= -(\lambda^2(\lambda + 1) + 2\lambda(\lambda + 1) + 3(\lambda + 1))$$
$$= -(\lambda + 1)(\lambda^2 + 2\lambda + 3),$$

whereupon $\det(A - \lambda I) = 0$ yields $\lambda_1 = -1$ and $\lambda_{2,3} = -1 \pm i\sqrt{2}$. We find that $\text{Re}(\lambda_{1,2,3}) < 0$. Hence, in view of Corollary 9.2, the trivial solution of the considered system is asymptotically stable.

Example 9.9. Consider the quasi-linear Stieltjes differential system

$$x'_{1g} = -2x_1 + x_2 + 3x_3 + 8x_1^3 + x_2^3,$$
$$x'_{2g} = -6x_2 - 5x_3 + 7x_3^4,$$
$$x'_{3g} = -x_3 + x_1^4 + x_2^2 + x_3^3, \quad t \in I.$$

Here,

$$A = \begin{pmatrix} -2 & 1 & 3 \\ 0 & -6 & -5 \\ 0 & 0 & -1 \end{pmatrix}, \quad f(t,x) = \begin{pmatrix} 8x_1^3 + x_2^3 \\ 7x_3^4 \\ x_1^4 + x_2^2 + x_3^3 \end{pmatrix}.$$

Then, $f(t,x) = o(||x||)$ uniformly in t as $||x|| \to 0$. Next,

$$\det(A - \lambda \mathscr{I}) = \det \begin{pmatrix} -2-\lambda & 1 & 3 \\ 0 & -6-\lambda & -5 \\ 0 & 0 & -1-\lambda \end{pmatrix}$$

$$= (-1-\lambda)(-2-\lambda)(-6-\lambda),$$

whereupon, $\det(A - \lambda \mathscr{I}) = 0$ yields $\lambda_1 = -2$, $\lambda_2 = -6$, and $\lambda_3 = -1$. We find that $\operatorname{Re}(\lambda_{1,2,3}) < 0$. Hence, in view of Corollary 9.2, the trivial solution of the considered system is asymptotically stable.

Exercise 9.6. Prove that the trivial solution of the quasi-linear Stieltjes differential system

$$x'_{1g} = -x_1 + x_1^4 + x_2^4,$$
$$x'_{2g} = x_1 - x_2 + x_1^2 + x_3^6,$$
$$x'_{3g} = x_1 + 5x_2 - 4x_3 + x_3^3, \quad t \in I,$$

is asymptotically stable.

Exercise 9.7. Suppose that the matrix A possesses at least one eigenvalue with a positive real part and the function f satisfies the condition (9.18). Prove that the trivial solution of the system (9.16) is unstable.

Exercise 9.8. Prove that the trivial solution of the quasi-linear Stieltjes differential system

$$x'_{1g} = x_1 + x_1^2 + x_2^2 + x_3^2,$$
$$x'_{2g} = -x_1 + 2x_2 + x_3^3,$$
$$x'_{3g} = -x_1 - x_2 - x_3 + x_2^2, \quad t \in I,$$

is unstable.

Now, we suppose that

$$\|f(t,x)\| \leq \lambda(t)\|x\|, \quad t \in I, \quad x \in \mathbb{R}^n, \tag{9.19}$$

where λ is a nonnegative continuous function on I such that

$$\int_{t_0}^{\infty} \lambda(s) d_g s < \infty.$$

Theorem 9.14. *Suppose that the solutions of the Stieltjes differential system*

$$x'_g = B(t)x, \quad t \in I, \tag{9.20}$$

are uniformly stable and the function f satisfies the condition (9.19). Then the trivial solution of the system (9.17) is uniformly stable.

Proof. Let Ψ be a fundamental matrix of the system (9.20). Since the solutions of the system (9.20) are uniformly stable, there is constant $c > 0$ such that

$$||\Psi(t)(\Psi(s))^{-1}|| \leq c, \quad t \geq s \geq t_0.$$

Let x be a solution of the system (9.17) such that

$$x(t_1) = x_1, \quad t_1 \geq t_0.$$

Then

$$x(t) = \Psi(t)(\Psi(t_1))^{-1}x_1 + \int_{t_1}^{t} \Psi(t)(\Psi(s))^{-1} f(s, x(s)) d_g s, \quad t \geq t_1.$$

Hence,

$$||x(t)|| = \left\| \Psi(t)(\Psi(t_1))^{-1}x_1 + \int_{t_1}^{t} \Psi(t)(\Psi(s))^{-1} f(s, x(s)) d_g s \right\|$$

$$\leq ||\Psi(t)(\Psi(t_1))^{-1}|| ||x_1|| + \int_{t_1}^{t} ||\Psi(xt)(\Psi(s))^{-1}||$$

$$||f(s, x(s))|| d_g s$$

$$\leq c||x_1|| + c \int_{t_1}^{t} \lambda(s) ||x(s)|| d_g s, \quad t \geq t_1.$$

From the last inequality and from the Stieltjes–Gronwall-type inequality,[1] it follows that

$$||x(t)|| \leq c||x_1|| \exp\left[c \int_{t_1}^{t} \lambda(s) d_g s\right]$$

$$\leq c||x_1|| \exp\left[c \int_{t_1}^{\infty} \lambda(s) d_g s\right], \quad t \geq t_1.$$

---

[1] As in the classical case, one can deduct a Stieltjes analogue of the classical Gronwall inequality. Since the proof repeats the steps in the classical case, we leave it to the reader for an exercise.

Let
$$M = c\exp\left[c\int_{t_1}^{\infty}\lambda(s)d_g s\right], \quad t \geq t_1.$$

Therefore for a given $\varepsilon > 0$, if $||x_1|| < \frac{\varepsilon}{M}$, then we obtain
$$||x(t)|| < \varepsilon, \quad t \geq t_1,$$

i.e., the trivial solution of (9.17) is uniformly stable. This completes the proof. □

Theorem 9.15. *Let the solutions of the system* (9.20) *be asymptotically stable and the function f satisfies the condition* (9.19). *Then the trivial solution of* (9.17) *is asymptotically stable.*

Proof. Let Ψ be a fundamental matrix of the system (9.20). Since the solutions of the system (9.20) are asymptotically stable, we have that
$$||\Psi(t)|| \to 0 \quad \text{as } t \to \infty.$$

Hence, for given $\varepsilon > 0$, there is $t_2 \in (t_0, \infty)$ such that
$$||\Psi(t)(\Psi(t_0))^{-1}x_0|| < \varepsilon, \quad t \geq t_2.$$

Also, there is a constant $c > 0$ such that
$$||\Psi(t)(\Psi(s))^{-1}|| \leq c, \quad t \geq s \geq t_0.$$

Hence, if x is a solution to the system (9.17) such that $x(t_0) = x_0$, we have
$$||x(t)|| = \left\|\Psi(t)(\Psi(t_0))^{-1}x_0 + \int_{t_0}^{t}\Psi(t)(\Psi(s))^{-1}f(s,x(s))d_g s\right\|$$
$$\leq ||\Psi(t)(\Psi(t_0))^{-1}||\,||x_0|| + \int_{t_0}^{t}||\Psi(t)(\Psi(s))^{-1}||\,||f(s,x(s))||d_g s$$
$$\leq \varepsilon + c\int_{t_0}^{t}||f(s,x(s))||d_g s$$
$$\leq \varepsilon + c\int_{t_0}^{t}\lambda(s)||x(s)||d_g s, \quad t \geq t_2.$$

From the last inequality and from the Stieltjes–Gronwall-type inequality, we get

$$\|x(t)\| \leq \varepsilon \exp\left[c \int_{t_0}^{\infty} \lambda(s) d_g s\right]$$

$$\leq \varepsilon \exp\left[c \int_{t_0}^{\infty} \lambda(s) d_g s\right].$$

Since $\varepsilon > 0$ was arbitrarily chosen and the fact that

$$\exp\left[c \int_{t_0}^{\infty} \lambda(s) d_g s\right]$$

does not depend on ε or t_2, we conclude that

$$\|x(t)\| \to 0 \quad \text{as } t \to \infty,$$

i.e., the trivial solution of the system (9.17) is asymptotically stable. This completes the proof. □

Example 9.10. Consider the Stieltjes differential equation

$$x'''_g - 2x''_g + 3x'_g + 9(\sin x)^2 = 0, \quad t \in I.$$

Let $x = x_1$, $x'_{1g} = x_2$, and $x'_{2g} = x_3$. Then,

$$x'_{3g} = x''_{2g}$$
$$= x'''_{1g}$$
$$= x'''_g$$
$$= 2x''_g - 3x'_g - 9(\sin x)^2$$
$$= -3x_2 + 2x_3 - 9(\sin x_1)^2, \quad t \in I,$$

Thus, we get the system

$$x'_{1g} = x_2,$$
$$x'_{2g} = x_3,$$
$$x'_{3g} = -3x_2 + 2x_3 - 9(\sin x_1)^2, \quad t \in I.$$

Here,

$$A = \begin{pmatrix} 0 & 1 & 0 \\ 0 & 0 & 1 \\ 0 & -3 & 2 \end{pmatrix} \quad \text{and} \quad f(t,x) = \begin{pmatrix} 0 \\ 0 \\ -9(\sin x_1)^2 \end{pmatrix}.$$

Then, $f(t,x) = o(||x||)$ uniformly in t as $||x|| \to 0$. Next,

$$\det(A - \lambda \mathscr{I}) = \det \begin{pmatrix} -\lambda & 1 & 0 \\ 0 & -\lambda & -1 \\ 0 & -3 & 2-\lambda \end{pmatrix}$$
$$= -\lambda^2(\lambda - 2) - 3\lambda$$
$$= -\lambda(\lambda^2 - 2\lambda + 3),$$

whereupon $\det(A - \lambda \mathscr{I}) = 0$ yields $\lambda_1 = 0$ and $\lambda_{2,3} = 1 \pm i\sqrt{2}$. We find that $\operatorname{Re}(\lambda_1) = 0$ and $\operatorname{Re}(\lambda_{2,3}) > 0$. Therefore, keeping in mind the assertion in Exercise 9.7, we can conclude that the trivial solution of the considered equation is unstable.

Exercise 9.9. Test for stability the trivial solution of the Stieltjes differential equation

$$x_g''' + 3x_g'' + 3x_g' + x + (\sin x)^2 = 0, \quad t \in I.$$

9.5 Two-Dimensional Autonomous Stieltjes Differential Systems

Hereafter, we consider the following Stieltjes differential system:

$$\begin{aligned} x_{1g}' &= f_1(x_1, x_2), \\ x_{2g}' &= f_2(x_1, x_2), \quad t \in I, \end{aligned} \tag{9.21}$$

where f_1 and f_2 together with their first-order Stieltjes partial derivatives are continuous in some domain D of the (x_1, x_2)-plane. Let $(x_{01}, x_{02}) \in D$. Then, the system (9.21) has a unique solution (x_1, x_2)

on some interval I containing the point t_0 such that

$$x_1(t_0) = x_{01},$$
$$x_2(t_0) = x_{02}.$$

Theorem 9.16. *Let (x_1, x_2) be a solution of the system (9.21) in the interval (α, β). Then for any constant $c \in \mathbb{R}$,*

$$x(t+c) = (x_1(t+c), x_2(t+c)), \quad t \in I,$$

is a solution to the system (9.21) which is defined in $(\alpha - c, \beta - c)$.

Proof. Since (x_1, x_2) is a solution of the system (9.21) which is defined in (α, β), we have that $(x_1(t+c), x_2(t+c))$ is defined in $(\alpha - c, \beta - c)$. Also,

$$x'_{1g}(t+c) = f_1(x_1(t+c), x_2(t+c)),$$
$$x'_{2g}(t+c) = f_2(x_1(x+c), x_2(x+c)), \quad t \in I.$$

This completes the proof. □

Definition 9.8. In the domain D of (x_1, x_2)-plane, any solution of the Stieltjes differential system (9.21) may be regarded as a parametrical curve given by $(x_1(t), x_2(t))$ with t as the parameter. The curve $(x_1(t), x_2(t))$ is called trajectory or an orbit, or a path of the Stieltjes differential system (9.21), and the (x_1, x_2)-plane is called the phase plane.

Remark 9.1. From Theorem 9.16, it follows that, for any constant c, both $(x_1(t), x_2(t))$, $x \in (\alpha, \beta)$ and $(x_1(t+c), x_2(t+c))$, $x \in (\alpha - c, \beta - c)$ represent the same trajectory.

Theorem 9.17. *Through each point $(x_{01}, x_{02}) \in D$, there passes one and only one trajectory of the system (9.21).*

Proof. Suppose that there are two different trajectories, $(x_1(t), x_2(t))$ and $(y_1(t), y_2(t))$, defined on the intervals I_1 and I_2, respectively, of the system (9.21) passing through the point (x_{01}, x_{02}).

Then, there are $t_0 \in I_1$ and $t_1 \in I_2$ such that
$$(x_1(t_0), x_2(t_0)) = (y_1(t_1), y_2(t_1))$$
$$= (x_{01}, x_{02}).$$
By the existence and uniqueness theorem, it follows that for $t_0 \neq t_1$, we note that $(x_{11}(t), x_{21}(t))$, where
$$x_{11}(t) = x_1(t - t_1 + t_0) \quad \text{and} \quad x_{21}(t) = x_2(t - t_1 + t_0),$$
is also a solution to the system (9.21). For this solution, we have
$$x_{11}(t_1) = x_1(t_0)$$
$$= x_{01}$$
$$= y_1(t_1),$$
and
$$x_{21}(t_1) = x_2(t_0)$$
$$= x_{02}$$
$$= y_2(t_1).$$
Hence, in view of the existence and uniqueness theorem, we conclude that
$$x_{11}(t) = y_1(t) \quad \text{and} \quad x_{21}(t) = y_2(t).$$
Since $(x_1(t), x_2(t))$ and $(x_{11}(t), x_{21}(t))$ represent the same trajectory, we conclude that $(x_1(t), x_2(t))$ and $(y_1(t), y_2(t))$ represent the same trajectory. This completes the proof. $\square$

Example 9.11. Consider the Stieltjes differential system
$$x'_{1g} = \frac{1}{x_2},$$
$$x'_{2g} = -\frac{1}{x_1}, \quad t \in I.$$
Then we find that
$$(x_1(t), x_2(t)) = (t + c, -t - c), \quad c \in \mathbb{R},$$
is an infinite number of solutions to the considered system. However, they represent the same trajectory
$$x_1 x_2 = 1.$$

Example 9.12. Consider Stieltjes differential system

$$x'_{1g} = \frac{2}{x_2^{\frac{1}{2}}},$$

$$x'_{2g} = -\frac{2}{x_1^3}, \quad t \in I.$$

Then we find that

$$(x_1(t), x_2(t)) = \left((t+c)^2, \frac{1}{(t+c)^2}\right), \quad c \in \mathbb{R},$$

is an infinite number of solutions of the considered system. However, they represent the same trajectory

$$x_1 x_2 = 1.$$

Example 9.13. Consider Stieltjes differential system

$$x'_{1g} = -2x_2,$$

$$x'_{2g} = 2x_1, \quad t \in I.$$

Then we find that

$$(x_1(t), x_2(t)) = (\cos(2t+c), \sin(2t+c)), \quad t \in I, \quad c \in [0, 2\pi),$$

is an infinite number of solutions of the considered system. However, they represent the same curve

$$x_1^2 + x_2^2 = 1.$$

Definition 9.9. A point $(x_{01}, x_{02}) \in D$ at which both f_1 and f_2 vanish simultaneously is called a critical point of the system (9.21).

Remark 9.2. The critical point is also known as the point of equilibrium or stationary point or singular point or rest point.

If (x_{01}, x_{02}) is a critical point of the system (9.21), then

$$x_1(x) = x_{01},$$

$$x_2(x) = x_{02}$$

is a solution to the system (9.21). Therefore, no trajectory can pass through the point (x_{01}, x_{02}).

Definition 9.10. A critical point (x_{01}, x_{02}) is said to be isolated provided there exists no other critical point in some neighbourhood of (x_{01}, x_{02}).

Let (x_{01}, x_{02}) be a critical point for the system (9.21). The substitution

$$y_1 = x_1 - x_{01},$$
$$y_2 = x_2 - x_{02}$$

transforms the system (9.21) into an equivalent system with $(0,0)$ as critical point. Therefore, without any restriction, we can assume that $(0,0)$ is a critical point of the system (9.21). We can rewrite the system (9.21) in the form

$$\begin{aligned} x'_{1g} &= a_{11}x_1 + a_{12}x_2 + h_1(x_1, x_2), \\ x'_{2g} &= a_{21}x_1 + a_{22}x_2 + h_2(x_1, x_2), \quad t \in I, \end{aligned} \quad (9.22)$$

where a_{11}, a_{12}, a_{21}, and a_{22} are some constants and

$$h_1(0,0) = h_2(0,0) = 0$$

and

$$\lim_{x_1, x_2 \to 0} \frac{h_1(x_1, x_2)}{(x_1^2 + x_2^2)^{\frac{1}{2}}} = \lim_{x_1, x_2 \to 0} \frac{h_2(x_1, x_2)}{(x_1^2 + x_2^2)^{\frac{1}{2}}} = 0.$$

Using the results of the previous sections, we get the following result.

Theorem 9.18.

(1) *If the zero solution of the system*

$$\begin{aligned} x'_{1g} &= a_{11}x_1 + a_{12}x_2, \\ x'_{2g} &= a_{21}x_1 + a_{22}x_2, \quad t \in I. \end{aligned} \quad (9.23)$$

is asymptotically stable, then the zero solution of the system (9.21) is asymptotically stable.
(2) *If the zero solution of the system (9.23) is unstable, then the zero solution of the system (9.21) is unstable.*
(3) *If the zero solution of the system (9.23) is stable, then the zero solution of the system (9.21) may be asymptotically stable, stable or unstable.*

Definition 9.11. The picture of all trajectories of the Stieltjes differential system is called the phase portrait of the system.

We note that the solutions of the system (9.23) can be determined explicitly. Therefore, a complete description of its phase portrait can be given. The nature of the solutions of the system (9.23) depends on the eigenvalues of the matrix

$$A = \begin{pmatrix} a_{11} & a_{12} \\ a_{21} & a_{22} \end{pmatrix}.$$

We have that

$$\det(A - \lambda \mathscr{I}) = \det \begin{pmatrix} a_{11} - \lambda & a_{12} \\ a_{21} & a_{22} - \lambda \end{pmatrix}$$

$$= (\lambda - a_{11})(\lambda - a_{12}) - a_{12}a_{21}$$

$$= \lambda^2 - (a_{11} + a_{22})\lambda + a_{11}a_{22} - a_{12}a_{21}$$

$$= \lambda^2 - (\operatorname{tr} A)\lambda + \det A.$$

Let λ_1 and λ_2 be the roots of the equation

$$\lambda^2 - (\operatorname{tr} A)\lambda + \det A = 0. \tag{9.24}$$

Case I: Let $\lambda_1, \lambda_2 \in \mathbb{R}$ be such that $\lambda_1 \neq \lambda_2$.

Let y_1 and y_2 be the eigenvectors of A corresponding to λ_1 and λ_2. Then the general solution of the system (9.23) can be written as

$$\begin{pmatrix} x_1(t) \\ x_2(t) \end{pmatrix} = c_1 \begin{pmatrix} y_{11} \\ y_{12} \end{pmatrix} e_{\lambda_1, g}(t, t_0) + c_2 \begin{pmatrix} y_{21} \\ y_{22} \end{pmatrix} e_{\lambda_2, g}(t, t_0), \quad t \in I, \tag{9.25}$$

where c_1 and c_2 are constants. Without any restriction, we suppose that $\lambda_1 > \lambda_2$.

(1) Suppose $\lambda_2 < \lambda_1 < 0$. Then the critical point $(0,0)$ is asymptotically stable.

- Assume that $c_1 = 0$ and $c_2 \neq 0$. In this case,
$$x_1(t) = c_2 y_{21} e_{\lambda_2, g}(t, t_0),$$
$$x_2(t) = c_2 y_{22} e_{\lambda_2, g}(t, t_0), \quad t \in I,$$

whereupon
$$\frac{x_2(t)}{x_1(t)} = \frac{y_{22}}{y_{21}}, \quad t \in I,$$

i.e.,
$$x_2(t) = \frac{y_{22}}{y_{21}} x_1(t), \quad t \in I,$$

Thus, the trajectory is a straight line with Stieltjes slope $\frac{y_{22}}{y_{21}}$.

- Assume that $c_1 \neq 0$ and $c_2 = 0$. In this case,
$$x_1(t) = c_1 y_{11} e_{\lambda_1, g}(t, t_0),$$
$$x_2(t) = c_1 y_{12} e_{\lambda_1, g}(t, t_0), \quad t \in I,$$

whereupon
$$\frac{x_2(t)}{x_1(t)} = \frac{y_{12}}{y_{11}}, \quad t \in I,$$

i.e.,
$$x_2(t) = \frac{y_{12}}{y_{11}} x_1(t), \quad t \in I,$$

i.e., the trajectory is a straight line with Stieltjes slope $\frac{y_{12}}{y_{11}}$.

- Assume that $c_1 \neq 0$ and $c_2 \neq 0$. In this case,
$$x_1(t) = c_1 y_{11} e_{\lambda_1, g}(t, t_0) + c_2 y_{21} e_{\lambda_2, g}(t, t_0),$$
$$x_2(t) = c_1 y_{12} e_{\lambda_1, g}(t, t_0) + c_2 y_{22} e_{\lambda_2, g}(t, t_0), \quad t \in I,$$

whereupon
$$\frac{x_2(t)}{x_1(t)} = \frac{c_1 y_{11} e_{\lambda_1, g}(t, t_0) + c_2 y_{21} e_{\lambda_2, g}(t, t_0)}{c_1 y_{12} e_{\lambda_1, g}(t, t_0) + c_2 y_{22} e_{\lambda_2, g}(t, t_0)}$$
$$= \frac{c_1 y_{12} + c_2 y_{22} e_{(\lambda_2 \ominus_g \lambda_1), g}(t, t_0)}{c_1 y_{11} + c_2 y_{21} e_{(\lambda_2 \ominus_g \lambda_1), g}(t, t_0)}, \quad t \in I. \quad (9.26)$$

From the last expression, we have that
$$\frac{x_2(t)}{x_1(t)} \to \frac{y_{12}}{y_{11}} \quad \text{as } t \to \infty$$
and
$$x_1(t) \to 0 \quad \text{and} \quad x_2(t) \to 0 \quad \text{as } t \to \infty.$$

Thus, all trajectories tend to $(0,0)$ with Stieltjes slope $\frac{y_{12}}{y_{11}}$. Also,
$$\frac{x_2(t)}{x_1(t)} = \frac{c_1 y_{12} e_{(\lambda_1 \ominus_g \lambda_2), g}(t, t_0) + c_2 y_{22}}{c_1 y_{11} e_{(\lambda_1 \ominus_g \lambda_2), g}(t, t_0) + c_2 y_{21}}, \quad t \in I. \qquad (9.27)$$

Hence,
$$\frac{x_2(t)}{x_1(t)} \to \frac{y_{22}}{y_{21}} \quad \text{as } t \to -\infty.$$

Thus, all trajectories become asymptotic to the line with Stieltjes slope $\frac{y_{22}}{y_{21}}$.

Definition 9.12. Here, the critical point is called a stable node.

(2) Suppose $\lambda_1 > \lambda_2 > 0$. Then all nontrivial solutions tend to ∞ as $t \to \infty$. Therefore the critical point $(0,0)$ is unstable. Using (9.26), we have that
$$\frac{x_2(t)}{x_1(t)} \to \frac{y_{12}}{y_{11}} \quad \text{as } t \to \infty,$$
i.e., the trajectories become asymptotic to the line with Stieltjes slope
$$\frac{y_{12}}{y_{11}}.$$

Using (9.27), we have
$$\frac{x_2(t)}{x_1(t)} \to \frac{y_{22}}{y_{21}} \quad \text{as } t \to -\infty$$
and
$$x_1(t) \to 0 \quad \text{and} \quad x_2(t) \to 0 \quad \text{as } t \to -\infty,$$

i.e., the trajectories tend to $(0,0)$ with Stieltjes slope

$$\frac{y_{22}}{y_{21}}$$

as $t \to -\infty$.

Definition 9.13. Here, the critical point is called unstable node.

Example 9.14. We consider the Stieltjes differential system

$$x'_{1g} = 4x_1 + x_2,$$
$$x'_{2g} = 3x_1 + 6x_2, \quad t \in I.$$

Here the critical point is $(0,0)$ and

$$A = \begin{pmatrix} 4 & 1 \\ 3 & 6 \end{pmatrix}.$$

Then

$$\det(A - \lambda \mathscr{I}) = \det \begin{pmatrix} 4 - \lambda & 1 \\ 3 & 6 - \lambda \end{pmatrix}$$
$$= (\lambda - 6)(\lambda - 4) - 3$$
$$= \lambda^2 - 10\lambda + 21,$$

whereupon $\det(A - \lambda \mathscr{I}) = 0$ yields $\lambda_1 = 7$ and $\lambda_2 = 3$. Hence, in view of Exercise 9.7, the trivial solution of the considered system is unstable. Thus, the critical point $(0,0)$ is unstable node.

Example 9.15. We consider the Stieltjes differential system

$$x'_{1g} = 3x_1 + x_2,$$
$$x'_{2g} = -x_1.$$

Here the critical point is $(0,0)$ and

$$A = \begin{pmatrix} 3 & 1 \\ -1 & 0 \end{pmatrix}.$$

Then
$$\det(A - \lambda \mathscr{I}) = \det\begin{pmatrix} 3 - \lambda & 1 \\ -1 & -\lambda \end{pmatrix}$$
$$= \lambda(\lambda - 3) + 1$$
$$= \lambda^2 - 3\lambda + e$$
$$= \lambda^2 - 3\lambda + 1,$$

whereupon $\det(A - \lambda \mathscr{I}) = 0$ yields $\lambda_{1,2} = \frac{3 \pm \sqrt{5}}{2}$. Hence, in view of Exercise 9.7, the trivial solution of the considered system is unstable. Thus, the critical point $(0,0)$ is unstable node.

Example 9.16. We consider the Stieltjes differential system
$$x'_{1g} = -x_1 + 2x_2,$$
$$x'_{2g} = -3x_2, \quad t \in I.$$

Here the critical point is $(0,0)$ and
$$A = \begin{pmatrix} -1 & 2 \\ 0 & -3 \end{pmatrix}.$$

Then
$$\det(A - \lambda \mathscr{I}) = \det\begin{pmatrix} -1 - \lambda & 2 \\ 0 & -3 - \lambda \end{pmatrix}$$
$$= (-1 - \lambda)(-3 - \lambda),$$

whereupon $\det(A - \lambda \mathscr{I}) = 0$ yields $\lambda_1 = -1$ and $\lambda_2 = -3$. Hence, in view of Corollary 9.1, the trivial solution of the considered system is stable. Thus, the critical point $(0,0)$ is stable node.

Exercise 9.10. Determine the type of stability of the critical point $(0,0)$ of each of the following linear Stieltjes differential systems:

(a)
$$x'_{1g} = 3x_1,$$
$$x'_{2g} = 2x_1 + x_2, \quad t \in I,$$

(b)
$$x'_{1g} = 3x_1 - x_2,$$
$$x'_{2g} = x_1, \quad t \in I,$$

(c)
$$x'_{1g} = -10x_1,$$
$$x'_{2g} = x_1 - 7x_2, \quad t \in I.$$

(3) Suppose $\lambda_1 > 0 > \lambda_2$. In this case, the general solution of the system (9.23) is given by (9.25).

- Assume that $c_1 = 0$ and $c_2 \neq 0$. Then, as in the previous case, we have
$$\frac{x_2(t)}{x_1(t)} = \frac{y_{22}}{y_{21}}, \quad t \in I,$$

i.e.,
$$x_2(t) = \frac{y_{22}}{y_{21}} x_1(t), \quad t \in I,$$

whereupon
$$x_1(t) \to 0 \quad \text{and} \quad x_2(t) \to 0 \quad \text{as } t \to \infty$$

or
$$x_1(t) \to \infty \quad \text{and} \quad x_2(t) \to \infty \quad \text{as } t \to -\infty.$$

- Assume that $c_1 \neq 0$ and $c_2 = 0$. Then, as in the previous case, we have
$$\frac{x_2(t)}{x_1(t)} = \frac{x_{12}}{y_{11}}, \quad t \in I,$$

i.e.,
$$x_2(t) = \frac{y_{12}}{y_{11}} x_1(t), \quad t \in I,$$

whereupon
$$x_1(t) \to \infty \quad \text{and} \quad x_2(t) \to \infty \quad \text{as } t \to \infty$$

or
$$x_1(t) \to 0 \quad \text{and} \quad x_2(t) \to 0 \quad \text{as } t \to -\infty.$$

- Assume that $c_1 \neq 0$ and $c_2 \neq 0$. Using (9.26), we get
$$\frac{x_2(t)}{x_1(t)} \to \frac{y_{12}}{y_{11}} \quad \text{as } t \to \infty.$$
Hence, all trajectories are asymptotic to the line with Stieltjes slope
$$\frac{y_{12}}{y_{11}}$$
as $t \to \infty$. Now, using (9.27), we find that all trajectories are asymptotic to the line with Stieltjes slope
$$\frac{y_{22}}{y_{21}}$$
as $t \to \infty$. In this case,
$$x_1(t) \to \infty \quad \text{and} \quad x_2(t) \to \infty \quad \text{as } t \to \pm\infty.$$

Definition 9.14. This type of critical point is called a saddle point.

The saddle point is an unstable critical point of the system.

Example 9.17. We consider the Stieltjes differential system
$$x'_{1g} = -x_1 + x_2,$$
$$x'_{2g} = -x_1 + 3x_2, \quad t \in I.$$
Here the critical point is $(0,0)$ and
$$A = \begin{pmatrix} -1 & 1 \\ 1 & 3 \end{pmatrix}.$$
Then
$$\det(A - \lambda \mathscr{I}) = \det \begin{pmatrix} -1-\lambda & 1 \\ -1 & 3-\lambda \end{pmatrix}$$
$$= (\lambda - 3)(\lambda + 1) + 1$$
$$= \lambda^2 - 2\lambda - 3 + 1$$
$$= \lambda^2 - 2\lambda - 2,$$
whereupon $\det(A - \lambda \mathscr{I}) = 0$ yields $\lambda_1 = 1 - \sqrt{3}$ and $\lambda_2 = 1 + \sqrt{3}$. Therefore, the critical point $(0,0)$ is a saddle point.

Exercise 9.11. Determine the type of stability of the critical point $(0,0)$ of the system
$$x'_{1g} = -3x_1 + x_2,$$
$$x'_{2g} = -x_1 + x_2, \quad t \in I.$$

Case II: Let $\lambda_1, \lambda_2 \in \mathbb{R}$ be such that $\lambda_1 = \lambda_2 = \lambda$.
In this case, the general solution of the system (9.23) is given by
$$\begin{pmatrix} x_1(t) \\ x_2(t) \end{pmatrix} = c_1 \begin{pmatrix} 1 + (a_{11} - \lambda)g(t) \\ a_{21}g)(t) \end{pmatrix} e_{\lambda,g}(t, t_0)$$
$$+ c_2 \begin{pmatrix} a_{12}g(t) \\ e + (a_{22} - \lambda)g(t) \end{pmatrix} e_{\lambda,g}(t, t_0),$$
where c_1 and c_2 are arbitrary constants.

(1) Suppose that $\lambda < 0$. Then, both $x_1(t)$ and $x_2(t)$ tend to 0 as $t \to \infty$ and hence, the critical point $(0,0)$ of the system (9.23) is asymptotically stable. Also,
$$\frac{x_2(t)}{x_1(t)} = \frac{c_2 + (a_{21}c_1 + (a_{22} - \lambda)c_2)g(t)}{c_1 + (a_{12}c_2 + (a_{11} - \lambda)c_1)g(t)}, \quad t \in I.$$

- Assume that $a_{11} = a_{22} \neq 0$ and $a_{12} = a_{21} = 0$. Then,
$$\lambda = a_{11}$$
$$= a_{22}$$

and
$$\frac{x_2(t)}{x_1(t)} = \frac{c_2}{c_1}.$$
Therefore, all trajectories are straight lines with the Stieltjes slope
$$\frac{c_2}{c_1}.$$

Definition 9.15. Here, the critical point is called as stable proper (star-shaped) node.

- Assume that $a_{11}, a_{12}, a_{21},$ and a_{22} are arbitrary real numbers. In this case,
$$\frac{x_1(t)}{x_2(t)} \to \frac{a_{21}c_1 + (a_{22} - \lambda)c_2}{a_{12}c_2 + (a_{11} - \lambda)c_1} \quad \text{as } t \to \pm\infty.$$

Since
$$(a_{11} - \lambda)(a_{22} - \lambda) = a_{12}a_{21},$$
we have
$$\frac{a_{11} - \lambda}{a_{12}} = \frac{a_{21}}{a_{22} - \lambda}.$$
Then,
$$\frac{a_{21}c_1 + (a_{22} - \lambda)c_2}{a_{12}c_2 + (a_{11} - \lambda)c_1} = \frac{a_{21}\left(c_1 + \frac{a_{22}-\lambda}{a_{21}}c_2\right)}{(a_{11} - \lambda)\left(\frac{a_{12}}{a_{11}-\lambda}\right)c_2 + c_1}$$
$$= \frac{a_{21}}{a_{11} - \lambda}.$$

Therefore, all trajectories are asymptotic to the line
$$x_1 = \frac{a_{21}}{(a_{11} - \lambda)}x_2 \quad \text{as } t \to \pm\infty.$$

Definition 9.16. Here, the origin $(0,0)$ is called stable improper node.

(2) Suppose $\lambda > 0$. In this case, all solutions tend to ∞ as $t \to \infty$. Therefore, the critical point $(0,0)$ of the system (9.23) is unstable. The trajectories are the same as for the case $\lambda < 0$ except that the direction of the motion is reversed.

Definition 9.17. The critical point $(0,0)$ of the system (9.23) is called unstable proper (star-shaped) node provided $a_{11} = a_{22} > 0$ and $a_{12} = a_{21} = 0$.

Definition 9.18. The critical point $(0,0)$ of the system (9.23) is called unstable improper node provided $a_{12} \neq 0$ or $a_{21} \neq 0$ and $\lambda_1 = \lambda_2 > 0$.

Example 9.18. We consider the Stieltjes differential system
$$x'_{1g} = -x_1,$$
$$x'_{2g} = -x_2, \quad t \in I.$$

Here the critical point is $(0,0)$ and
$$A = \begin{pmatrix} -1 & 0 \\ 0 & -1 \end{pmatrix}$$
and
$$a_{11} = a_{22}$$
$$= -1,$$
$$a_{12} = a_{21}$$
$$= 0.$$
Also, the roots of the equation $\det(A - \lambda \mathscr{I}) = 0$ are $\lambda_1 = \lambda_2 = -1$. Therefore, the critical point $(0,0)$ is stable proper (star-shaped) node.

Example 9.19. We consider the Stieltjes differential system
$$x'_{1g} = -3x_1 + x_2,$$
$$x'_{2g} = -x_1 - x_2, \quad t \in I.$$
Here the critical point is $(0,0)$ and
$$A = \begin{pmatrix} -3 & 1 \\ -1 & -1 \end{pmatrix}$$
and
$$a_{11} = -3,$$
$$a_{12} = 1,$$
$$a_{21} = -1,$$
$$a_{22} = -1.$$
Then
$$\det(A - \lambda \mathscr{I}) = \det \begin{pmatrix} -3 - \lambda & 1 \\ -1 & -1 - \lambda \end{pmatrix}$$
$$= (\lambda + 1)(\lambda + 3) + 1$$
$$= \lambda^2 + 4\lambda + 3 + 1$$
$$= \lambda^2 + 4\lambda + 4$$
$$= (\lambda + 2)^2,$$

whereupon $\det(A - \lambda \mathscr{I}) = 0$ yields $\lambda_1 = \lambda_2 = -2 < 0$. Therefore, the critical point $(0,0)$ is stable improper node.

Example 9.20. We consider the Stieltjes differential system

$$x'_{1g} = 2x_1,$$
$$x'_{2g} = 2x_2, \quad t \in I.$$

Here the critical point is $(0,0)$ and

$$A = \begin{pmatrix} 2 & 0 \\ 0 & 2 \end{pmatrix}$$

and

$$a_{11} = a_{22}$$
$$= 2,$$
$$a_{12} = a_{21}$$
$$= 0.$$

Also, the roots of the equation $\det(A - \lambda \mathscr{I}) = 0$ are $\lambda_1 = \lambda_2 = 2$. Therefore, the critical point $(0,0)$ is unstable proper (star-shaped) node.

Example 9.21. We consider the Stieltjes differential system

$$x'_{1g} = 2x_1 - x_2,$$
$$x'_{2g} = x_1 + 4x_2, \quad t \in I.$$

Here the critical point is $(0,0)$ and

$$A = \begin{pmatrix} 2 & -1 \\ 1 & 4 \end{pmatrix}$$

$$a_{11} = 2,$$
$$a_{12} = -1,$$
$$a_{21} = 0,$$
$$a_{22} = 4.$$

Then
$$\det(A - \lambda \mathscr{I}) = \det \begin{pmatrix} -\lambda & -1 \\ 1 & 4-\lambda \end{pmatrix}$$
$$= (\lambda - 4)(\lambda - 2) + 1$$
$$= \lambda^2 - 6\lambda + 8 + 1$$
$$= \lambda^2 - 6\lambda + 9$$
$$= (\lambda - 3)^2,$$

whereupon $\det(A - \lambda I) = 0$ yields $\lambda_1 = \lambda_2 = 3$. Therefore, the critical point $(0,0)$ is unstable improper node.

Exercise 9.12. Determine the type of stability of the critical point $(0,0)$ of the following Stieltjes differential systems:

(1)
$$x'_{1g} = -4x_1,$$
$$x'_{2g} = -4x_2, \quad t \in I,$$

(2)
$$x'_{1g} = -2x_1 - 3x_2,$$
$$x'_{2g} = 2x_1 - 6x_2, \quad t \in I,$$

(3)
$$x'_{1g} = \frac{1}{2}x_1,$$
$$x'_{2g} = \frac{1}{2}x_2, \quad t \in I,$$

(4)
$$x'_{1g} = 2x_1 - 3x_2,$$
$$x'_{2g} = 2x_1 + 6x_2 \quad t \in I.$$

Case III: Let λ_1 and λ_2 be complex conjugates.
Let $\lambda_1 = \alpha + i\beta$ and $\lambda_2 = \alpha - i\beta$. Without any restriction, we assume that $\beta > 0$.

(1) Suppose $\alpha = 0$. In this case, the critical point $(0,0)$ is stable, but not asymptotically stable.

Definition 9.19. Here, the critical point $(0,0)$ is called center.

(2) Suppose $\alpha < 0$. In this case, the critical point is asymptotically stable.

Definition 9.20. Here, the critical point $(0,0)$ is called stable focus.

(3) Suppose $\alpha > 0$. In this case, the critical point $(0,0)$ is unstable.

Definition 9.21. Here, the critical point $(0,0)$ is called unstable focus.

Example 9.22. We consider the Stieltjes differential system

$$x'_{1g} = -x_1 - 2x_2,$$
$$x'_{2g} = x_1 - x_2 \quad t \in I.$$

Here the critical point is $(0,0)$ and

$$A = \begin{pmatrix} -1 & -2 \\ 1 & -1 \end{pmatrix}.$$

Then

$$\det(A - \lambda \mathscr{I}) = \det \begin{pmatrix} -1 - \lambda & -2 \\ 1 & -1 - \lambda \end{pmatrix}$$
$$= (\lambda + 1)^2 + 2$$
$$= \lambda^2 + 2\lambda + 3,$$

whereupon $\det(A - \lambda \mathscr{I}) = 0$ yields $\lambda_{1,2} = -1 \pm \sqrt{2}i$. Since $\operatorname{Re}(\lambda_{1,2}) = -1 < 0$, by Definition 9.20, we conclude that the critical point $(0,0)$ is stable focus.

Example 9.23. We consider the Stieltjes differential system
$$x'_{1g} = 2x_1 + x_2,$$
$$x'_{2g} = -4x_1 + 2x_2.$$
Here the critical point is $(0,0)$ and
$$A = \begin{pmatrix} 2 & 1 \\ -4 & 2 \end{pmatrix}.$$
Then
$$\det(A - \lambda \mathscr{I}) = \det \begin{pmatrix} 2-\lambda & 1 \\ -4 & 2-\lambda \end{pmatrix}$$
$$= (\lambda - 2)^2 + 4$$
$$= \lambda^2 - 4\lambda + 8$$
$$= \lambda^2 - 4\lambda + 8,$$
whereupon $\det(A - \lambda \mathscr{I}) = 0$ yields $\lambda_{1,2} = 2 \pm 2i$. Since $\text{Re}(\lambda_{1,2}) = 2 > 0$, by Definition 9.21, we conclude that the critical point $(0,0)$ is unstable focus.

Example 9.24. We consider the Stieltjes differential system
$$x'_{1g} = x_2,$$
$$x'_{2g} = -9x_1, \quad t \in I.$$
Here the critical point is $(0,0)$ and
$$A = \begin{pmatrix} 0 & 1 \\ -9 & 0 \end{pmatrix}.$$
Then
$$\det(A - \lambda \mathscr{I}) = \det \begin{pmatrix} -\lambda & 1 \\ -9 & -\lambda \end{pmatrix}$$
$$= \lambda^2 + 9$$
$$= \lambda^2 + 9,$$
whereupon $\det(A - \lambda \mathscr{I}) = 0$ yields $\lambda_{1,2} = \pm 3i$. Since $\text{Re}(\lambda_{1,2}) = 0$, by definition 9.19, we conclude that the critical point $(0,0)$ is center.

Exercise 9.13. Determine the critical point $(1,1)$ of the following Stieltjes differential systems on I:

(1)
$$x'_{1g} = 3x_2,$$
$$x'_{2g} = -12x_1, \quad t \in I,$$

(2)
$$x'_{1g} = x_1 + x_2,$$
$$x'_{2g} = -5x_1 + x_2, \quad t \in I,$$

(3)
$$x'_{1g} = -2x_1 + 3x_2,$$
$$x'_{2g} = -7x_1 - 2x_2, \quad t \in I.$$

Example 9.25. We consider the Stieltjes differential system
$$x'_{1g} = x_1 + 4x_2,$$
$$x'_{2g} = x_1 + x_2 - x_1^2, \quad t \in I. \tag{9.28}$$

The critical points satisfy the system
$$x_1 + 4x_2 = 0,$$
$$x_1 + x_2 - x_1^2 = 0.$$

Solving these systems simultaneously for x_1 and x_2, we obtain $x_1 = 0$ and $x_2 = 0$ or $x_1 = 3/4$ and $x_2 = -3/16$. Thus, the system (9.28) has two critical points: $(0,0)$ and $(3/4, -3/16)$.

Case I: Consider the critical point $(0,0)$.

The associated linear Stieltjes differential system corresponding to the system (9.28) is
$$x'_{1g} = x_1 + 4x_2,$$
$$x'_{2g} = x_1 + x_2, \quad t \in I.$$

Here
$$A = \begin{pmatrix} 1 & 4 \\ 1 & 1 \end{pmatrix}.$$

Then

$$\det(A - \lambda \mathscr{I}) = \det \begin{pmatrix} 1-\lambda & 4 \\ 1 & 1-\lambda \end{pmatrix}$$
$$= (\lambda - 1)^2 - 4$$
$$= (\lambda - 3)(\lambda + 1),$$

whereupon $\det(A - \lambda \mathscr{I}) = 0$ yields $\lambda_1 = 3$ and $\lambda_2 = -1$. Therefore, the critical $(0,0)$ is a saddle point.

Case II: Consider the critical point $\left(\frac{3}{4}, \frac{-3}{16}\right)$.

We substitute

$$y_1 = x_1 - \frac{3}{4}$$

and

$$y_2 = x_2 + \frac{3}{16}.$$

Then we obtain

$$y'_{1g} = x'_{1g}$$

and

$$y'_{2g} = x'_{2g},$$

and

$$x_1 = y_1 + \frac{3}{4},$$
$$x_2 = y_2 - \frac{3}{16}.$$

In this case, the system (9.28) takes the form

$$y'_{1g} = y_1 + \frac{3}{4} + 4\left(y_2 - \frac{3}{16}\right),$$
$$y'_{2g} = y_1 + \frac{3}{4} + y_2 - \frac{3}{16} - \left(y_1 + \frac{3}{4}\right)^2,$$

i.e.,
$$y'_{1g} = y_1 + 4y_2,$$
$$y'_{2g} = -\frac{1}{2}y_1 + y_2 - y_1^2.$$

The associated linear Stieltjes differential system corresponding to the above system is
$$y'_{1g} = y_1 + 4y_2,$$
$$y'_{2g} = -\frac{1}{2}y_1 + y_2.$$

Here
$$A = \begin{pmatrix} 1 & 4 \\ -\frac{1}{2} & 1 \end{pmatrix}.$$

Then
$$\det(A - \lambda \mathscr{I}) = \det \begin{pmatrix} 1-\lambda & 4 \\ -\frac{1}{2} & 1-\lambda \end{pmatrix}$$
$$= (\lambda - 1)^2 + 2$$
$$= \lambda^2 - 2\lambda + 3,$$

whereupon $\det(A - \lambda \mathscr{I}) = 0$ yields $\lambda_{1,2} = 1 \pm \sqrt{2}i$. Since $\mathrm{Re}(\lambda_{1,2}) = 1 > 0$, the critical point $(3/4, -3/16)$ is unstable focus.

Example 9.26. We consider the Stieltjes differential system
$$x'_{1g} = x_1 - x_2 + x_1^2,$$
$$x'_{2g} = 12x_1 - 6x_2 + x_1 x_2, \quad t \in I. \tag{9.29}$$

The critical points of (9.29) satisfy the system
$$x_1 - x_2 + x_1^2 = 0,$$
$$12x_1 - 6x_2 + x_1 x_2 = 0.$$

Solving this system simultaneously for x_1 and x_2, we obtain $x_1 = 0$ and $x_2 = 0$ or $x_1 = 2$ and $x_2 = 6$ or $x_1 = 3$ and $x_2 = 12$. Thus, the system (9.29) has three critical points: $(0,0)$, $(2,6)$, and $(3,12)$.

Case I: Consider the critical point $(0,0)$.

The associated linear Stieltjes differential system corresponding to the system (9.29) is

$$x'_{1g} = x_1 - x_2,$$
$$x'_{2g} = 12x_1 - 6x_2, \quad t \in I.$$

Here

$$A = \begin{pmatrix} 1 & -1 \\ 12 & -6 \end{pmatrix}.$$

Then

$$\det(A - \lambda \mathscr{I}) = \det \begin{pmatrix} 1 - \lambda & -1 \\ 12 & -6 - \lambda \end{pmatrix}$$
$$= (\lambda + 6)(\lambda - 1) + 12$$
$$= \lambda^2 + 5\lambda + 6,$$

whereupon $\det(A - \lambda \mathscr{I}) = 0$ yields $\lambda_1 = -2$ and $\lambda_2 = -3$. Hence, in view of Corollary 9.1, the trivial solution of the considered system is stable. Thus, the critical point $(0,0)$ is stable node.

Case II: Consider the critical point $(2,6)$.

We substitute

$$z_1 = x_1 - 2$$

and

$$z_2 = x_2 - 6.$$

Then we obtain

$$x_1 = y_1 + 2,$$
$$x_2 = y_2 + 6,$$

and

$$x'_{1g} = y'_{1g},$$
$$x'_{2g} = y'_{2g}.$$

In this case, the system (9.29) takes the form
$$y'_{1g} = y_1 + 2 - y_2 - 6 + (y_1 + 2)^2,$$
$$y'_{2g} = 12(y_1 + 2) - 6(y_2 + 6) + (y_1 + 2)(y_1 + 6),$$
i.e.,
$$y'_{1g} = 5y_1 - y_2 + y_1^2,$$
$$y'_{2g} = 20y_1 - 6y_2 + y_1^2.$$
The associated linear Stieltjes differential system corresponding to the above system is
$$y'_{1g} = 5y_1 - y_2,$$
$$y'_{2g} = 20y_1 - 6y_2.$$
Here
$$A = \begin{pmatrix} 5 & -1 \\ 20 & -6 \end{pmatrix}.$$
Then
$$\det(A - \lambda \mathscr{I}) = \det \begin{pmatrix} 5-\lambda & -1 \\ 20 & -6-\lambda \end{pmatrix}$$
$$= (\lambda + 6)(\lambda - 5) + 20$$
$$= \lambda^2 + \lambda - 10,$$
whereupon $\det(A - \lambda \mathscr{I}) = 0$ yields $\lambda_1 = \frac{-1+\sqrt{41}}{2}$ and $\lambda_2 = \frac{-1-\sqrt{41}}{2}$. Therefore, $(2, 6)$ is saddle point.

Case III: Consider the critical point $(3, 12)$.

We substitute
$$y_1 = x_1 - 3$$
and
$$y_2 = x_2 - 12.$$
Then we obtain
$$x_1 = y_1 + 3,$$
$$x_2 = y_2 + 12,$$

and
$$x'_{1g} = y'_{1g},$$
$$x'_{2g} = y'_{2g}.$$

In this case, the system (9.29) takes the form
$$y'_{1g} = y_1 + 3 - y_2 - 12 + (y_1 + 3)^2,$$
$$y'_{2g} = 12(y_1 + 3) - 6(y_2 + 12) + (y_1 + 3)(y_2 + 12),$$

i.e.,
$$y'_{1g} = 7y_1 - y_2 + y_1^2,$$
$$y'_{2g} = 24y_1 - 3y_2 + y_1 y_2.$$

The associated linear Stieltjes differential system corresponding to the above system is
$$y'_{1g} = 7y_1 - y_2,$$
$$y'_{2g} = 24y_1 - 3y_2.$$

Here
$$A = \begin{pmatrix} 7 & -1 \\ 24 & -3 \end{pmatrix}.$$

Then
$$\det(A - \lambda \mathscr{I}) = \det \begin{pmatrix} 7 - \lambda & -1 \\ 24 & -3 - \lambda \end{pmatrix}$$
$$= (\lambda + 3)(\lambda - 7) + 24$$
$$= \lambda^2 - 4\lambda + 3,$$

whereupon $\det(A - \lambda \mathscr{I}) = 0$ yields $\lambda_1 = 1$ and $\lambda_2 = 3$. Consequently, the critical point $(3, 12)$ is unstable node.

Example 9.27. We consider the Stieltjes differential system
$$x'_{1g} = x_1 - x_1^3 - x_1 x_2^2,$$
$$x'_{2g} = 2x_2 - x_2^5 - x_1^4 x_2, \quad t \in I. \tag{9.30}$$

The critical points of (9.30) satisfy the system
$$x_1 - y_1^3 - x_1 x_2^2 = 0,$$
$$2x_2 - x_2^5 - x_1^4 x_2 = 0.$$
Solving this system simultaneously for x_1 and x_2, we obtain $y_1 = 0$ and $x_2 = 0$ or $x_1 = 1$ and $x_2 = 0$ or $x_1 = -1$ and $x_2 = 0$. Thus, the system (9.30) has three critical points: $(0,0)$, $(1,0)$, and $(-1,0)$.

Case I: Consider the critical point $(0,0)$.

The associated linear Stieltjes differential system corresponding to the system (9.30) is
$$x'_{1g} = x_1,$$
$$x'_{2g} = 2x_2, \quad t \in I.$$

Here
$$A = \begin{pmatrix} 1 & 0 \\ 0 & 2 \end{pmatrix}.$$

Then
$$\det(A - \lambda \mathscr{I}) = \det \begin{pmatrix} 1-\lambda & 0 \\ 0 & 2-\lambda \end{pmatrix}$$
$$= (1-\lambda)(2-\lambda),$$

whereupon $\det(A - \lambda \mathscr{I}) = 0$ yields $\lambda_1 = 1$ and $\lambda_2 = 2$. Consequently, the critical point $(0,0)$ is unstable node.

Case II: Consider the critical point $(1,0)$.

We substitute $y_1 = x_1 - 1$, whereupon
$$x_1 = y_1 + 1$$
and
$$x'_{1g} = y'_{1g}.$$

In this case, the system (9.30) takes the form
$$y'_{1g} = y_1 + 1 - (y_1 + 1)^3 - (y_1 + 1)x_2^2,$$
$$x'_{2g} = 2x_2 - x_2^5 - (y_1 + 1)^4 x_2,$$

i.e.,
$$y'_{1g} = -2y_1 - y_1^3 - 3y_1^2 - y_1x_2^2 - x_2^2,$$
$$x'_{2g} = x_2 - x_2^5 - (y_1^4 + 4y_1^3 + 6y_1^2 + 4y_1)x_2.$$

The associated linear Stieltjes differential system corresponding to the above system is
$$y'_{1g} = -2y_1,$$
$$x'_{2g} = x_2.$$

Here
$$A = \begin{pmatrix} -2 & 0 \\ 0 & 1 \end{pmatrix}.$$

Then
$$\det(A - \lambda \mathscr{I}) = \det \begin{pmatrix} -2-\lambda & 0 \\ 0 & 1-\lambda \end{pmatrix}$$
$$= (1-\lambda)(-2-\lambda),$$

whereupon $\det(A - \lambda \mathscr{I}) = 0$ yields $\lambda_1 = 1$ and $\lambda_2 = -2$. Consequently, the critical point $(1,0)$ is saddle point.

Case III: Consider the critical point $(-1, 0)$.

We substitute $y_1 = x_1 + 1$, whereupon
$$x_1 = y_1 - 1$$

and
$$x'_{1g} = y'_{1g}.$$

In this case, the system (9.30) takes the form
$$y'_{1g} = y_1 - 1 - (y_1 - 1)^3 - (y_1 - 1)x_2^2,$$
$$y'_{2g} = -2x_2 - x_2^5 - (y_1 - 1)^4 x_2,$$

i.e.,
$$y'_{1g} = -2y_1 - y_1^3 + 3y_1^2 - (y_1 - 1)x_2^2,$$
$$x'_{2g} = -x_2 - x_2^5 - (y_1^4 - 4y_1^3 + 6y_1^2 - 4y_1)x_2.$$

The associated linear Stieltjes differential system corresponding to the above system is
$$y'_{1g} = -2y_1,$$
$$x'_{2g} = -x_2.$$

Here
$$A = \begin{pmatrix} -2 & 0 \\ 0 & -1 \end{pmatrix}.$$

Then
$$\det(A - \lambda \mathscr{I}) = \det \begin{pmatrix} -2 - \lambda & 0 \\ 0 & -1 - \lambda \end{pmatrix}$$
$$= (-1 - \lambda)(-2 - \lambda)$$
$$= (1 + \lambda)(2 + \lambda),$$

whereupon $\det(A - \lambda \mathscr{I}) = 0$ yields $\lambda_1 = -1$ and $\lambda_2 = -2$. Consequently, the critical point $(-1, 0)$ is stable node.

Exercise 9.14. Find the critical points of the following Stieltjes differential systems and determine their nature on I:

(1)
$$x'_{1g} = 2x_1 + x_2^2 - 1,$$
$$x'_{2g} = 6x_1 - x_2^2 + 1.$$

(2)
$$x'_{1g} = x_2^2 - 4x_1^2,$$
$$x'_{2g} = 4x_2 - 8.$$

(3)
$$x'_{1g} = 4 - 4x_1 - 2x_2,$$
$$x'_{2g} = x_1 x_2.$$

(4)
$$x'_{1g} = 2 + x_2 - x_1^2,$$
$$x'_{2g} = 2x_1(x_1 - x_2).$$

(5)
$$x'_{1g} = x_1 x_2 - 4,$$
$$x'_{2g} = (x_1 - 4)(x_2 - x_1).$$

(6)
$$x'_{1g} = 1 - x_1^2 - x_2^2,$$
$$x'_{2g} = 2x_1 x_2.$$

Exercise 9.15. Consider the second-order Stieltjes differential equation
$$x''_g + \mu(x^2 - 1)x'_g + x = 0 \quad t \in I. \tag{9.31}$$
where μ is a constant. This equation is called as the Stieltjes–Van der Pol equation.

(1) Prove that the (9.31) is equivalent to the system
$$\begin{aligned} x'_{1g} &= x_2, \\ x'_{2g} &= -x_1 + \mu(1 - x_1^2)x_2 \quad t \in I. \end{aligned} \tag{9.32}$$

(2) Find the critical points of (9.32) and determine their nature.

Solution. (1) Let $x_1 = x$ and $x_2 = x'_g$. Then, $x'_{1g} = x_2$ and
$$\begin{aligned} x'_{2g} &= x''_g \\ &= -\mu(x^2 - 1)x'_g - x \\ &= -\mu(x_1^2 - 1)x_2 - x_1 \\ &= -x_1 + \mu x_2 - \mu x_1^2 x_2, \end{aligned}$$
i.e.,
$$\begin{aligned} x'_{1g} &= x_2, \\ x'_{2g} &= -x_1 + \mu x_2 - \mu x_1^2 x_2. \end{aligned}$$
which is essentially (9.32).

Now, we consider the system (9.32). We differentiate the first equation in the sense of Stieltjes and get

$$x''_{1g} = x'_{2g}$$
$$= -x_1 + \mu x_2 - \mu x_1^2 x_2$$
$$= -x_1 + \mu x'_{1g} - \mu x_1^2 x'_{1g},$$

which is essentially (9.31).

(2) The critical points of (9.31) satisfy the system

$$x_2 = 0,$$
$$-x_1 + \mu x_2 - \mu x_1^2 x_2 = 0.$$

This gives

$$x_1 = 0,$$
$$x_2 = 0,$$

i.e., the system (9.32) has only one critical point $(0,0)$. The associated linear Stieltjes differential system corresponding to the system (9.32) is

$$x'_{1g} = x_2,$$
$$x'_{2g} = -x_1 + \mu x_2.$$

Here

$$A = \begin{pmatrix} 0 & 1 \\ -1 & \mu \end{pmatrix}.$$

Then

$$\det(A - \lambda \mathscr{I}) = \det \begin{pmatrix} -\lambda & 1 \\ -1 & \mu - \lambda \end{pmatrix}$$
$$= \lambda(\lambda - \mu) + 1$$
$$= \lambda^2 - \mu\lambda + 1$$
$$= \lambda^2 - \mu\lambda + 1,$$

whereupon $\det(A - \lambda \mathscr{I}) = 0$ yields

$$\lambda_1 = \frac{\mu + \sqrt{\mu^2 - 4}}{2},$$

$$\lambda_2 = \frac{\mu - \sqrt{\mu^2 - 4}}{2}.$$

Now, depending on the value of μ, we have the following three cases:

Case I: $\mu \in (-\infty, -2) \cup (2, +\infty)$.

In this case, we have $\mu^2 - 4 > 0$.

(a) Suppose $\lambda_1 > \lambda_2 > 0$. Then

$$\mu + \sqrt{\mu^2 - 4} > 0,$$
$$\mu - \sqrt{\mu^2 - 4} > 0,$$

and

$$\sqrt{\mu^2 - 4} > -\mu,$$
$$\sqrt{\mu^2 - 4} < \mu,$$

whereupon $\mu > 0$. Consequently, if $\mu \in (2, \infty)$, then the critical point $(0,0)$ is unstable node.

(b) Suppose $0 > \lambda_1 > \lambda_2$. Then

$$\mu + \sqrt{\mu^2 - 4} < 0,$$
$$\mu - \sqrt{\mu^2 - 4} < 0,$$

and

$$\sqrt{\mu^2 - 4} < -\mu,$$
$$\sqrt{\mu^2 - 4} > \mu,$$

whereupon

$$\mu < 0.$$

Consequently, if $\mu \in (-\infty, -2)$, then the critical point $(0,0)$ is stable node.

(c) Suppose $\lambda_1 > 0 > \lambda_2$. Then
$$\mu + \sqrt{\mu^2 - 4} > 0,$$
$$\mu - \sqrt{\mu^2 - 4} < 0,$$
and
$$\sqrt{\mu^2 - 4} > -\mu,$$
$$\sqrt{\mu^2 - 4} > \mu.$$

- For $\mu \in (2, +\infty)$, we get
$$\lambda_1 > 0 > \lambda_2$$
and
$$\sqrt{\mu^2 - 4} > \mu,$$
which is impossible.
- For $\mu \in (-\infty, -2)$, we get
$$0 > \lambda_1 > \lambda_2$$
and
$$\sqrt{\mu^2 - 4} > -\mu,$$
which is impossible.

Case II: $\mu^2 - 4 = 0$ or $\mu = \pm 2$.
Then
$$\lambda_1 = \lambda_2 = \mu.$$
If $\mu = -2$, then the critical point $(0,0)$ is stable improper node.
If $\mu = 2$, then the critical point $(0,0)$ is unstable improper node.

Case III: $\mu^2 - 4 < 0$ or $\mu \in (-2, 2)$.
Then
$$\lambda_1 = \frac{\mu + \sqrt{4 - \mu^2}\, i}{2}$$

and
$$\lambda_2 = \frac{\mu - \sqrt{4-\mu^2}i}{2}.$$

If $\mu \in (-2, 0)$, then the critical point $(0,0)$ is stable focus. If $\mu = 0$, then the critical point $(0,0)$ is center. If $\mu \in (0, 2)$, then the critical point $(0,0)$ is unstable focus. □

Exercise 9.16. Consider the second-order Stieltjes differential equation

$$x_g'' + \mu \left(\frac{1}{3}(x_g')^2 - 1\right)x_g' + x = 0 \quad t \in I, \tag{9.33}$$

where μ is a constant. This equation is called as the Stieltjes–Rayleigh equation. Reduce (9.33) to (9.31).

Solution. Differentiating (9.33) in the sense of Stieltjes, we find

$$x_g''' + \mu \left(\frac{2}{3}x_g' x_g''\right)x_g' + \mu \left(\frac{1}{3}(x_g')^2 - 1\right)x_g'' + x_g' = 0$$

or

$$x_g''' + \mu \left(\frac{2}{3}(x_g')^2 + \frac{1}{3}(x_g')^2 - 1\right)x_g'' + x_g' = 0,$$

i.e.,

$$x_g''' + \mu \left((x_g')^2 - 1\right)x_g'' + x_g' = 0. \tag{9.34}$$

We substitute $x_g' = y$, whereupon

$$x_g'' = y_g',$$
$$x_g''' = y_g''$$

and (9.34) reduces to

$$y_g'' + \mu(y^2 - 1)y_g' + y = 0,$$

which is (9.31), the Stieltjes–Van der Pol equation. □

Exercise 9.17. Reduce the following Stieltjes differential equation:

$$x_g'' + px_g' + qx = 0, \quad t \in I.$$

where p and q are constants, to the Stieltjes differential system. Prove that its critical point $(0,0)$ is

(1) unstable node provided $p^2 > 4q$, $q > 0$, and $p < 0$,
(2) stable node provided $p^2 > 4q$, $q > 0$, and $p > 0$,
(3) saddle point provided $p^2 > 4q$ and $q < 0$,
(4) stable node provided $p^2 = 4q$ and $p > 0$,
(5) stable focus provided $p^2 < 4q$ and $p > 0$,
(6) unstable focus provided $p^2 < 4q$ and $p < 0$,
(7) stable center provided $p = 0$ and $q > 0$.

9.6 Advanced Practical Problems

Problem 9.1. Using the definition for stable/unstable solution, investigate the solutions of the following SIVPs on I:

(1)
$$x_g' = -t^3 x + 1, \quad t \in [0, \infty),$$
$$x(0) = 0.$$

(2)
$$x_g' = (t^2 + e)x + t, \quad t \in [0, \infty),$$
$$x(0) = 1.$$

(3)
$$x_g' = t^4, \quad t \in [0, \infty),$$
$$x(0) = 1.$$

Stability Theory for Stieltjes Differential Systems 331

Problem 9.2. Prove that all solutions of the Stieltjes differential system

$$x'_{1g} = -x_1 + 20x_2,$$
$$x'_{2g} = -2x_1 - x_2, \quad t \in I,$$

are stable.

Problem 9.3. Prove that all solutions of the Stieltjes differential system

$$x'_{1g} = -10x_1,$$
$$x'_{2g} = 2x_1 - x_2,$$
$$x'_{3g} = -3x_1 + 4x_2 - 17x_2, \quad t \in I,$$

are asymptotically stable.

Problem 9.4. Use the Hurwitz criteria to find the real parameter a such that the trivial solution of the Stieltjes differential system

$$x'_{1g} = x_2,$$
$$x'_{2g} = x_3,$$
$$x'_{3g} = ax_1 - 3x_2 - 2x_2, \quad t \in I,$$

is asymptotically stable.

Answer 9.2. $-6 < a < 0$.

Problem 9.5. Prove that the trivial solution of the Stieltjes differential system

$$x'_{1g} = -3x_1 + 4x_1^4 + 2x_2^3,$$
$$x'_{2g} = -x_2 + 4x_3 + 7x_3^4,$$
$$x'_{3g} = -4x_3 + x_1^4 + x_2^4 + x_3^4, \quad t \in I,$$

is asymptotically stable.

Problem 9.6. Prove that the trivial solution of the Stieltjes differential system

$$x'_{1g} = 2x_1 + x_2 - x_3^3,$$
$$x'_{2g} = -x_2 + x_3 + x_2^2,$$
$$x'_{3g} = -x_3 + x_1^2, \quad t \in I,$$

is unstable.

Problem 9.7. Test for stability the trivial solution of the Stieltjes differential equation

$$x'''_g + 3x''_g - 4x + \cos x - 1 = 0, \quad t \in I.$$

Problem 9.8. Determine the type of stability of the critical point $(0,0)$ of each of the following linear Stieltjes differential systems on I:

(1)
$$x'_{1g} = 2x_1 + x_2,$$
$$x'_{2g} = 2x_1 + 3x_2.$$

(2)
$$x'_{1g} = x_1 + x_2,$$
$$x'_{2g} = x_1 + 2x_2.$$

(3)
$$x'_{1g} = -2x_1 - 3x_2,$$
$$x'_{2g} = -4x_2.$$

Problem 9.9. Determine the type of stability of the critical point $(0,0)$ of the Stieltjes differential system on I:

$$x'_{1g} = -2x_1 - x_2,$$
$$x'_{2g} = x_1 + x_2.$$

Problem 9.10. Determine the type of stability of the critical point of the following Stieltjes differential systems on I:

(1)
$$x'_{1g} = -6x_1,$$
$$x'_{2g} = -6x_2.$$

(2)
$$x'_{1g} = -2x_1 - 3x_2,$$
$$x'_{2g} = 3x_1 - 8x_2.$$

(3)
$$x'_{1g} = 5x_1,$$
$$x'_{2g} = 5x_2.$$

(4)
$$x'_{1g} = 2x_1 + 8x_2,$$
$$x'_{2g} = -2x_1 + 10x_2.$$

Problem 9.11. Determine the critical point of the following Stieltjes differential systems on I:

(1)
$$x'_{1g} = 4x_2,$$
$$x'_{2g} = -10x_1.$$

(2)
$$x'_{1g} = 2x_1 - 7x_2$$
$$x'_{2g} = x_1 + 2x_2.$$

(3)
$$x'_{1g} = -3x_1 - 10x_2$$
$$x'_{2g} = x_1 - 3x_2.$$

Problem 9.12. Find the critical points of the following Stieltjes differential systems and determine their nature on I:

(1)
$$x'_{1g} = 2(x_1 - e)(x_2 - 2),$$
$$x'_{2g} = x_2^2 - x_1^2.$$

(2)
$$x'_{1g} = (x_1 + x_2)^2 - e,$$
$$x'_{2g} = -x_2^2 - x_1 + e.$$

(3)
$$x'_{1g} = (2x_1 - x_2)^2 - 9,$$
$$x'_{2g} = 9 - (x_1 - 2x_2)^2.$$

(4)
$$x'_{1g} = (2x_1 - x_2)^2 - 9,$$
$$x'_{2g} = (x_1 - 2x_2)^2 - 9.$$

(5)
$$x'_{1g} = x_1^2 + x_2^2 - 6x_1 - 8x_2,$$
$$x'_{2g} = x_1(2x_2 - x_1 + 5).$$

(6)
$$x'_{1g} = x_1^2 - x_2,$$
$$x'_{2g} = (x_1 - x_2)(x_1 - x_2 + 2).$$

Chapter 10

Linear Stieltjes Boundary Value Problems

The purpose of this chapter is to study Stieltjes differential equations with various types of boundary conditions. We investigate the existence and uniqueness of solutions of boundary value problems for both homogeneous and nonhomogeneous Stieltjes differential equations.

Suppose that $I \subseteq \mathbb{R}$ and $g \colon I \to \mathbb{R}$ is a monotone nondecreasing function that is continuous from the left everywhere.

10.1 Introduction

Let $\alpha, \beta \in I$ be such that $\alpha < \beta$ and $J = [\alpha, \beta]$. Consider the Stieltjes differential equation

$$p_0(t) x_g'' + p_1(t) x_g' + p_2(t) x = r(t), \quad t \in J, \tag{10.1}$$

where $p_0, p_1, p_2, r \in \mathscr{C}(J)$ are given functions. Together with (10.1), we consider the boundary conditions

$$\begin{aligned} l_1(x) &= a_0 x(\alpha) + a_1 x_g'(\alpha) + b_0 x(\beta) + b_1 x_g'(\beta) = a, \\ l_2(x) &= c_0 x(\alpha) + c_1 x_g'(\alpha) + d_0 x(\beta) + d_1 x_g'(\beta) = b, \end{aligned} \tag{10.2}$$

where $a_0, a_1, b_0, b_1, c_0, c_1, d_0, d_1, a, b \in \mathbb{R}$ are the given Stieltjes constants.

Definition 10.1. Equation (10.1) together with the boundary conditions (10.2) is said to be nonhomogeneous Stieltjes boundary value problem (SBVP).

Definition 10.2. The Stieltjes differential equation

$$p_0(t)x_g'' + p_1(t)x_g' + p_2(t)x = 0, \quad t \in J, \tag{10.3}$$

subject to the boundary conditions

$$\begin{aligned} l_1(x) &= 0, \\ l_2(x) &= 0, \end{aligned} \tag{10.4}$$

is said to be the homogeneous Stieltjes boundary value problem (SBVP).

The Stieltjes boundary conditions (10.2) include the following cases:

(1) Dirichlet–Stieltjes boundary conditions (first Stieltjes boundary condition)

$$\begin{aligned} x(\alpha) &= a, \\ x(\beta) &= b. \end{aligned}$$

(2) mixed Stieltjes boundary conditions (second Stieltjes boundary conditions)

$$\begin{aligned} x(\alpha) &= a, \\ x_g'(\beta) &= b, \end{aligned}$$

or

$$\begin{aligned} x_g'(\alpha) &= a, \\ x(\beta) &= b. \end{aligned}$$

(3) separated Stieltjes boundary conditions

$$\begin{aligned} a_0 x(\alpha) + a_1 x_g'(\alpha) &= a, \\ d_0 x(\beta) + d_1 x_g'(\beta) &= b, \end{aligned}$$

where

$$a_0^2 + a_1^2 \neq 0$$

and
$$d_0^2 + d_1^2 \neq 0.$$

(4) periodic Stieltjes boundary conditions
$$x(\alpha) = x(\beta),$$
$$x'_g(\alpha) = x'_g(\beta).$$

Definition 10.3. The SBVPs (10.1)–(10.2) and (10.3)–(10.4) are said to be regular Stieltjes boundary value problems provided $\alpha > 0$, $\beta < \infty$ and
$$p_0(x) \neq 0, \quad x \in J.$$
If $\alpha = 0$ or $\beta = \infty$, or $p_0(t_1) = 0$ for some $t_1 \in J$, the SBVPs (10.1)–(10.2) and (10.3)–(10.4) are said to be singular Stieltjes boundary value problems.

Theorem 10.1. *The maps* $l_1, l_2 \colon \mathscr{C}^1(J) \to \mathbb{R}$ *defined in* (10.2) *are linear Stieltjes maps.*

Proof. Let $x_1, x_2 \in \mathscr{C}^1(J)$ and $\gamma \in \mathbb{R}$ be arbitrarily chosen. Then
$$\begin{aligned}l_1(x_1 + x_2) &= a_0(x_1 + x_2)(\alpha) + a_1(x_1 + x_2)'_g(\alpha) + b_0(x_1 + x_2)(\beta) \\&\quad + b_1(x_1 + x_2)'_g(\beta) \\&= a_0(x_1(\alpha) + x_2(\alpha)) + a_1(x'_{1g}(\alpha) + x'_{2g}(\alpha)) + b_0(x_1(\beta) \\&\quad + x_2(\beta)) + b_1(x'_{1g}(\beta) + x'_{2g}(\beta)) \\&= a_0 x_1(\alpha) + a_1 x'_{1g}(\alpha) + b_0 x_1(\beta) + b_1 x'_{1g}(\beta) + a_0 x_2(\alpha) \\&\quad + a_1 x'_{2g}(\alpha) + b_0 x_2(\beta) + b_1 x'_{2g}(\beta) \\&= l_1(x_1) + l_2(x_2)\end{aligned}$$
and
$$\begin{aligned}l_1(\gamma x_1) &= a_0(\gamma x_1)(\alpha) + a_1(\gamma x_1)'_g(\alpha) + b_0(\gamma x_1)(\beta) + b_1(\gamma x_1)'_g(\beta) \\&= a_0 \gamma x_1(\alpha) + a_1 \gamma x'_{1g}(\alpha) + b_0 \gamma x_1(\beta) + b_1 \gamma x'_g(\beta) \\&= \gamma(a_0 x_1(\alpha) + a_1 x'_{1g}(\alpha) + b_0 x_1(\beta) + b_1 x'_{1g}(\beta)) \\&= \gamma l_1(x_1).\end{aligned}$$

Thus, $l_1\colon \mathscr{C}(J) \to \mathbb{R}$ is a linear map. The proof that $l_2\colon \mathscr{C}(J) \to \mathbb{R}$ is a linear map we leave to the reader as an exercise. This completes the proof. □

Definition 10.4. By a solution of SBVP (10.1)–(10.2), we mean a solution of (10.1) that satisfies the boundary conditions (10.2).

Definition 10.5. By a solution of SBVP (10.3)–(10.4), we mean a solution of (10.3) that satisfies the boundary conditions (10.4).

Suppose we know two linearly independent solutions x_1 and x_2 of (10.1). We will search the general solution of the SBVP (10.1)–(10.2) in the form
$$x(t) = k_1 x_1(t) + k_2 x_2(t), \quad t \in [\alpha, \beta],$$
where k_1 and k_2 are constants which will be determined by the system
$$a = a_0(k_1 x_1(\alpha) + k_2 x_2(\alpha)) + a_1(k_1 x'_{1g}(\alpha) + k_2 x'_{2g}(\alpha))$$
$$+ b_0(k_1 x_1(\beta) + k_2 x_2(\beta)) + b_1(k_1 x'_{1g}(\beta) + k_2 x'_{2g}(\beta)),$$
$$b = c_0(k_1 x_1(\alpha) + k_2 x_2(\alpha)) + c_1(k_1 x'_{1g}(\alpha) + k_2 x'_{2g}(\alpha))$$
$$+ d_0(k_1 x_1(\beta) + k_2 x_2(\beta)) + d_1(k_1 x'_{1g}(\beta) + k_2 x'_{2g}(\beta)).$$

Example 10.1. Consider the SBVP
$$x''_g + x = 0, \quad t \in [0, 1],$$
$$x(1) + x'_g(0) = 10,$$
$$x(0) + 3x'_g(1) = 4.$$

First, we will find a fundamental system of solutions of the given Stieltjes differential equation. Its characteristic equation is
$$r^2 + 1 = 0.$$
The roots of this equation are
$$r_{1,2} = \pm i.$$
Hence,
$$x_1(t) = \cos_{1,g}(t, 0),$$
$$x_2(t) = \sin_{1,g}(t, 0), \quad t \in [0, 1],$$

is a fundamental system of solutions for the considered Stieltjes differential equation. We will search the solution of the considered SBVP in the form

$$x(t) = a_1 \cos_{1,g}(t,0) + a_2 \sin_{1,g}(t,0), \quad t \in [0,1].$$

Here, a_1 and a_2 are constants. We have

$$x'_g(t) = -a_1 \sin_{1,g}(t,0) + a_2 \cos_{1,g}(t,0), \quad t \in [0,1],$$

and

$$x(0) = a_1 \cos_{1,g}(0,0) + a_2 \sin_{1,g}(0,0)$$
$$= a_1,$$
$$x(1) = a_1 \cos_{1,g}(1,0) + a_2 \sin_{1,g}(1,0),$$
$$x'_g(0) = -a_1 \sin_{1,g}(0,0) + a_2 \cos_{1,g}(0,0)$$
$$= a_2,$$
$$x'_g(1) = -a_1 \sin_{1,g}(1,0) + a_2 \cos_{1,g}(1,0).$$

Hence,

$$10 = x(0) + x'_g(0)$$
$$= a_1 + a_2,$$
$$4 = x(1) + 3x'_g(1)$$
$$= a_1 \cos_{1,g}(1,0) + a_2 \sin_{1,g}(1,0)$$
$$\quad + 3\left(-a_1 \sin_{1,g}(1,0) + a_2 \cos_{1,g}(1,0)\right)$$
$$= (a_1 + 3a_2) \cos_{1,g}(1,0) + (a_2 - 3a_1) \sin_{1,g}(1,0),$$

i.e., we get the system

$$10 = a_1 + a_2,$$
$$4 = (a_1 + 3a_2) \cos_{1,g}(1,0) + (a_2 - 3a_1) \sin_{1,g}(1,0),$$

whereupon

$$a_2 = 10 - a_1$$

and
$$\begin{aligned}4 &= (a_1 + 3(10 - a_1))\cos_{1,g}(1,0) + (10 - a_1 - 3a_1)\sin_{1,g}(1,0) \\ &= (a_1 - 3a_1 + 30)\cos_{1,g}(1,0) + (10 - 4a_1)\sin_{1,g}(1,0) \\ &= (-2a_1 + 30)\cos_{1,g}(1,0) + (10 - n4a_1)\sin_{1,g}(1,0) \\ &= -2(\cos_{1,g}(1,0) + 2\sin_{1,g}(1,0))a_1 \\ &\quad + 10(3\cos_{1,g}(1,0) + \sin_{1,g}(1,0)).\end{aligned}$$

This yields
$$\begin{aligned}2(\cos_{1,g}(1,0) &+ 2\sin_{1,g}(1,0))a_1 \\ &= 2\left(5(3\cos_{1,g}(1,0) + \sin_{1,g}(1,0)) - 2\right),\end{aligned}$$

i.e.,
$$a_1 = \frac{(5(3\cos_{1,g}(1,0) + \sin_{1,g}(1,0)) - 2)}{\cos_{1,g}(1,0) + 2\sin_{1,g}(1,0)}.$$

Next,
$$\begin{aligned}a_2 &= 10 - a_1 \\ &= 10 - \frac{(5(3\cos_{1,g}(1,0) + \sin_{1,g}(1,0)) - 2)}{\cos_{1,g}(1,0) + 2\sin_{1,g}(1,0)} \\ &= \frac{10\cos_{1,g}(1,0) + 20\sin_{1,g}(1,0) - 15\cos_{1,g}(1,0) - 5\sin_{1,g}(1,0) + 2}{\cos_{1,g}(1,0) + 2\sin_{1,g}(1,0)} \\ &= \frac{15\sin_{1,g}(1,0) - 5\cos_{1,g}(1,0) + 2}{\cos_{1,g}(1,0) + 2\sin_{1,g}(1,0)} \\ &= \frac{5(3\sin_{1,g}(1,0) - \cos_{1,g}(1,0) + 2)}{\cos_{1,g}(1,0) + 2\sin_{1,g}(1,0)}.\end{aligned}$$

Therefore,
$$\begin{aligned}x(t) &= \frac{(5(3\cos_{1,g}(1,0) + \sin_{1,g}(1,0)) - 2)}{\cos_{1,g}(1,0) + 2\sin_{1,g}(1,0)}\cos_{1,g}(t,0) \\ &\quad + \frac{5(3\sin_{1,g}(1,0) - \cos_{1,g}(1,0) + 2)}{\cos_{1,g}(1,0) + 2\sin_{1,g}(1,0)}\sin_{1,g}(t,0), \quad t \in [0,1],\end{aligned}$$

is the solution of the considered SBVP.

Exercise 10.1. Solve the following SBVP:
$$x_g'' x_g' - 2x = 0, \quad t \in [0, \infty),$$
$$x_g'(0) = 2,$$
$$x(\infty) = 1.$$

10.2 Existence of Solutions

Theorem 10.2. *Let x_1 and x_2 be two linearly independent solutions of (10.1). Then, the SBVP (10.3)–(10.4) has only the trivial solution if and only if*
$$\det \begin{pmatrix} l_1(x_1) & l_1(x_2) \\ l_2(x_1) & l_2(x_2) \end{pmatrix} \neq 0. \tag{10.5}$$

Proof. Since x_1 and x_2 are linearly independent solutions of (10.1), any solution x of (10.3) can be expressed in the form
$$x(t) = k_1 x_1(t) + k_2 x_2(t), \quad t \in J,$$
where k_1 and k_2 are constants. Hence, x satisfies (10.4) if and only if
$$l_1(k_1 x_1 + k_2 x_2) = l_1(k_1 x_1) + l_1(k_2 x_2)$$
$$= k_1 l_1(x_1) + k_2 l_1(x_2)$$
$$= 0$$

and

$$l_2(k_1 x_1 + k_2 x_2) = l_2(k_1 x_1) + l_2(k_2 x_2)$$
$$= k_1 l_2(x_1) + k_2 l_2(x_2)$$
$$= 0,$$

equivalently, k_1 and k_2 satisfy the system
$$k_1 l_1(x_1) + k_2 l_1(x_2) = 0,$$
$$k_1 l_2(x_1) + k_2 l_2(x_2) = 0,$$

equivalently,
$$c_1 = 0,$$
$$c_2 = 0$$

because (10.5) holds. This completes the proof. □

Corollary 10.1. *The SBVP* (10.3)–(10.4) *has infinite many solutions if and only if*

$$\det \begin{pmatrix} l_1(x_1) & l_1(x_2) \\ l_2(x_1) & l_2(x_2) \end{pmatrix} = 0. \qquad (10.6)$$

Example 10.2. Consider the SBVP

$$g(t)x'' - x'_g - 4(g(t))^3 x = 0, \quad t \in [1,2],$$
$$x(1) = 0,$$
$$x(2) = 0.$$

First, we will check that the functions

$$x_1(t) = \cosh_{(g(t))^2-1,g}(t,1),$$
$$x_2(t) = \frac{1}{2}\sinh_{(g(t))^2-1,g}(t,1), \quad t \in [1,2],$$

are solutions of the considered Stieltjes differential equation. We have

$$x'_{1g}(t) = 2g(t)\sinh_{(g(t))^2-1,g}(t,1),$$
$$x''_{1g}(t) = 2\sinh(g(t))^2 - 1, g(t,1) + 4(g(t))^2 \cosh_{(g(t))^2-1,g}(t,0),$$
$$x'_{2g}(t) = \frac{5}{2}g(t)\cosh_{(g(t))^2-1,g}(t,1),$$
$$x''_{2g}(t) = \frac{5}{2}\cosh_{(g(t))^2-1,g}(t,1)$$
$$\qquad + \frac{9}{2}(g(t))^2 \sinh_{(g(t))^2-1,g}(t,1), \quad t \in [1,2].$$

Then

$$g(t)x_{1g} +''(t) - x'_{1g}(t) - 4(g(t))^3 x_1(t)$$
$$= g(t)\left(2\sinh_{(g(t))^2-1,g}(t,1) + 4(g(t))^2 \cosh_{(g(t))^2-1,g}(t,1)\right)$$
$$\quad - 2g(t)\sinh_{(g(t))^2-1,g}(t,1) - 4(g(t))^3 \cosh_{(g(t))^2-1,g}(t,1)$$
$$= 0, \quad t \in [1,2],$$

and
$$g(t)x''_{2g}(t) - x'_{2g}(t) - 4(g(t))^3 x_2(t)$$
$$= g(t)\left(\frac{5}{2}\cosh_{(g(t))^2-1,g}(t,1) + \frac{9}{2}(g(t))^2 \sinh_{(g(t))^2-1,g}(t,1)\right)$$
$$- \frac{5}{2}g(t)\cosh_{(g(t))^2-1,g}(t,1) - 4(g(t))^3 \frac{1}{2}\sinh_{(g(t))^2-1,g}(t,1)$$
$$= 0, \quad t \in [1,2].$$

Therefore, x_1 and x_2 are the solutions of the considered Stieltjes differential equation. Next,

$$W_g(x_1, x_2)(t) = \det \begin{pmatrix} x_1(t) & x_2(t) \\ x'_{1g}(t) & x'_{2g}(t) \end{pmatrix}$$
$$= x_1(t)x'_{2g}(t) - x'_{1g}(t)x_2(t)$$
$$= \cosh_{(g(t))^2-1,g}(t,1)\frac{5}{2}g(t)\cosh_{(g(t))^2-1,g}(t,1)$$
$$- \frac{1}{2}g(t)t\sinh_{(g(t))^2-1,g}(t,1)\frac{1}{2}\sinh_{(g(t))^2-1,g}(t,1)$$
$$= \frac{5}{2}g(t)\left((\cosh_{(g(t))^2-1,g}(t,1))^2 - (\sinh_{(g(t))^2-1,g}(t,1))^2\right)$$
$$= \frac{5}{2}g(t)e_{(g(t))^2-1,g}(t,1)e_{-(g(t))^2-1,g}(t,1)$$
$$\neq 0, \quad t \in [1,2].$$

Consequently, x_1 and x_2 form a fundamental system of solutions of the considered Stieltjes differential equation. Here,
$$l_1(x) = x(1),$$
$$l_2(x) = x(2), \quad x \in \mathscr{C}_g^1([1,2]).$$
Then,
$$l_1(x_1) = x_1(1)$$
$$= \cosh_{2-1,g}(1,1)$$
$$= \cosh_{1,g}(1,1)$$
$$= 1,$$

$$l_1(x_2) = x_2(1)$$
$$= \frac{1}{2}\sinh_{1-1,g}(1,1)$$
$$= \frac{1}{2}\sinh_{1,g}(1,1)$$
$$= 0,$$
$$l_2(x_1) = x_1(2)$$
$$= \cosh_{4-1,g}(2,1)$$
$$= \cosh_{3,g}(2,1),$$

and

$$l_2(x_2) = x_2(2)$$
$$= \frac{1}{2}\sinh_{4-e,g}(2,1)$$
$$= \frac{1}{2}\sinh_{3,g}(2,1).$$

From here,

$$\det\begin{pmatrix} l_1(x_1) & l_1(x_2) \\ l_2(x_1) & l_2(x_2) \end{pmatrix} = \det\begin{pmatrix} 1 & 0 \\ \cosh_{3,g}(2,1) & \frac{1}{2}\sinh_{3,g}(2,1) \end{pmatrix}$$
$$= \frac{1}{2}\sinh_{3,g}(2,1)$$
$$\neq 0.$$

Consequently, in view of Theorem 10.2, the considered SBVP has only the trivial solution.

Exercise 10.2. Prove that the SBVP

$$g(t)x''_g - x'_g - 4t^3 x = 1, \quad t \in [1,2],$$
$$x'_g(1) = 0,$$
$$x'_g(2) = 0,$$

has only the trivial solution.

Linear Stieltjes Boundary Value Problems

Theorem 10.3. *The SBVP* (10.1)–(10.2) *has a unique solution if and only if the SBVP* (10.3)–(10.4) *has only the trivial solution.*

Proof. Let x_1 and x_2 form a fundamental system of solutions for the Stieltjes differential equation (10.3). Also, let x be a particular solution of (10.1). Then, the general solution of (10.1) has the representation

$$x(t) = k_1 x_1(t) + k_2 x_2(t) + y(t), \quad t \in J.$$

Here, k_1 and k_2 are constants. By the boundary conditions (10.2), we obtain the system

$$k_1 l_1(x_1) + k_2 l_1(x_2) + l_1(y) = a,$$
$$k_1 l_2(x_1) + k_2 m l_2(x_2) + l_2(y) = b.$$

This system has a unique solution (k_1, k_2) if and only if (10.5) holds. But, in view of Theorems 10.2, (10.5) holds if and only if the SBVP (10.3)–(10.4) has only the trivial solution. This completes the proof. $\square$

Exercise 10.3. Prove that the SBVP

$$g(t)x''_g - x'_g - 4(g(t))^3 x = 1 + 4(g(t))^4, \quad t \in [1, 2],$$
$$x(1) = 0,$$
$$x(2) = 1$$

has a unique solution.

10.3 Advanced Practical Problems

Problem 10.1. Solve the following SBVPs:

(1)
$$x''_g + 4x'_g + 7x = 0, \quad t \in [0, 1],$$
$$x(0) = 0,$$
$$x'_g(1) = 1,$$

(2)
$$x''_g - 6x'_g + 25x = 0, \quad t \in \left[0, \frac{\pi}{4}\right],$$
$$x'_g(0) = 1,$$
$$x\left(\frac{\pi}{4}\right) = 1,$$

(3)
$$(g(t))^2 x''_g + 7g(t)x'_g + 3x = 0, \quad t \in [1, 2],$$
$$x(1) = 1,$$
$$x(2) = 2,$$

(4)
$$x''_g + x = (g(t))^2, \quad t \in \left[0, \frac{\pi}{2}\right],$$
$$x(0) = 0,$$
$$x\left(\frac{\pi}{2}\right) = 1,$$

(5)
$$x''_g + 2x'_g + x = 0, \quad t \in [0, 2],$$
$$x(0) = 0,$$
$$x(2) = 3,$$

(6)
$$x''_g + x'_g + x = g(t), \quad t \in [0, 1],$$
$$x(0) + 2x'_g(0) = 1,$$
$$x(1) - x'_g(1) = 8,$$

Problem 10.2. Prove that the SBVP
$$x''_g + 2x'_g + 5x = 1, \quad t \in \left[0, \frac{\pi}{2}\right],$$
$$x(0) = 0,$$
$$x\left(\frac{\pi}{2}\right) = 1$$
has infinite number of solutions.

Problem 10.3. Prove that the SBVP

$$x''_g + x = g(t), \quad t \in \left[0, \frac{\pi}{2}\right],$$

$$x(0) + x'_g(0) = 1,$$

$$x\left(\frac{\pi}{2}\right) - x'_g\left(\frac{\pi}{2}\right) = \frac{\pi}{2}$$

has no solutions.

Problem 10.4. Let x_1 be a solution of the SBVP (10.1)–(10.4) and x_2 be a solution of the SBVP (10.3)–(10.2). Then, prove that

$$x = x_1 + x_2$$

is a solution to the SBVP (10.1)–(10.4).

Bibliography

Albés, I. M. and Tojo, F. A. F. Existence and uniqueness of solution for Stieltjes differential equations with several derivators. *Mediterranean Journal of Mathematics* 18: 1–31 (2021).

Bohner, M. and Peterson, A. *Dynamic Equations on Time Scales: An Introduction with Applications.* Springer Science & Business Media, New York, 2001.

Chen, Y., O'Regan, D. and Wang, J. R. Existence and stability of solutions for linear and nonlinear Stieltjes differential equations. *Quaestiones Mathematicae* 43(11): 1613–1638 (2020).

Fernández, F. J., Albés, I. M. and Tojo, F. A. F. Consequences of the product rule in Stieltjes differentiability (2022a). *arXiv preprint* arXiv:2205.10090.

Fernández, F. J., Albés, I. M. and Tojo, F. A. F. The Wronskian and the variation of parameters method in the theory of linear Stieltjes differential equations of second order (2022b). *arXiv preprint* arXiv:2206.10855.

Frigon, M. and Pouso, R. L. Theory and applications of first-order systems of Stieltjes differential equations. *Advances in Nonlinear Analysis* 6(1): 13–36 (2017).

Frigon, M. and Tojo, F. A. F. Stieltjes differential systems with nonmonotonic derivators. *Boundary Value Problems* 2020(1): 1–24 (2020).

Kim, Y.-J. Stieltjes derivatives and its applications to integral inequalities of Stieltjes type. *The Pure and Applied Mathematics* 18(1): 63–78 (2011).

Kloeden, P. E. and Panadiwal, J. Taylor expansions for continuous Stieltjes differential equations. *Bulletin of the Australian Mathematical Society* 48(2): 325–336 (1993).

Lopez Pouso, R. and Marquez Albes, I. Systems of Stieltjes differential equations with several derivators. *Mediterranean Journal of Mathematics* 16: 1–17 (2019a).

Lopez Pouso, R. and Marquez Albes, I. Resolution methods for mathematical models based on differential equations with Stieltjes derivatives. *Electronic Journal of Qualitative Theory of Differential Equations* 2019(72): 1–15 (2019b).

López Pouso, R. and Márquez Albés, I. Existence of extremal solutions for discontinuous Stieltjes differential equations. *Journal of Inequalities and Applications* 2020(1): 1–21 (2020).

Lopez Pouso, R., Marquez Albes, I. and Monteiro, G. Extremal solutions of systems of measure differential equations and applications in the study of Stieltjes differential problems. *Electronic Journal of Qualitative Theory of Differential Equations* 2018(38): 1–24 (2018).

Lopez Pouso, R., Márquez Albés, I. and Rodríguez-López, J. Solvability of non-semicontinuous systems of Stieltjes differential inclusions and equations. *Advances in Difference Equations* 2020(1): 1–14 (2020).

Marraffa, V. and Satco, B. Stieltjes differential inclusions with periodic boundary conditions without upper semicontinuity. *Mathematics* 10(1): 55 (2021).

Marquez Albes, I. Notes on the linear equation with Stieltjes derivatives. *Electronic Journal of Qualitative Theory of Differential Equations* 2021(42): 1–18 (2021).

Marquez Albes, I. and Monteiro, G. A. Notes on the existence and uniqueness of solutions of Stieltjes differential equations. *Mathematische Nachrichten* 294(4): 794–814 (2021).

Monteiro, G. and Satco, B. Distributional, differential and integral problems: Equivalence and existence results. *Electronic Journal of Qualitative Theory of Differential Equations* 2017(7): 1–26 (2017).

Monteiro, G. A. and Satco, B. Extremal solutions for measure differential inclusions via Stieltjes derivatives. *Advances in Difference Equations* 2019(1): 1–18 (2019).

Monteiro, G. A., Slavík, A. and Tvrdý, M. *Kurzweil–Stieltjes Integral: Theory and Applications*, World Scientific Publishing Company, Singapore, 2019.

Pouso, R. L. and Albés, I. M. General existence principles for Stieltjes differential equations with applications to mathematical biology. *Journal of Differential Equations* 264(8): 5388–5407 (2018).

Pouso, R. L. and Rodríguez, A. A new unification of continuous, discrete and impulsive calculus through Stieltjes derivatives. *Real Analysis Exchange* 40(2): 319–354 (2015).

Satco, B. and Smyrlis, G. Applications of Stieltjes derivatives to periodic boundary value inclusions. *Mathematics* 8(12): 2142 (2020a).

Satco, B. and Smyrlis, G. Periodic boundary value problems involving Stieltjes derivatives. *Journal of Fixed Point Theory and Applications* 22(4): 94 (2020b).

Schwabik, S. *Generalized Ordinary Differential Equations*, Vol. 5. World Scientific Publishing Company, Singapore, 1992.

Slavík, A. Dynamic equations on time scales and generalized ordinary differential equations. *Journal of Mathematical Analysis and Applications* 385(1): 534–550 (2012).

Index

$D_{[a,b]}$, 78
g, 101
$\mathscr{L}_g(f)(z)$, 121
$\mathscr{R}_g$, 91, 195
$\mathscr{R}_g(I, \mathbb{R}^{n \times n})$, 195
$\mathscr{R}_g(I)$, 195
$\ominus_g$, 96
$\oplus_g$, 91
$\cosh_{f,g}$, 114
$\cos_{f,g}$, 112
$\sinh_{f,g}$, 114
$\sin_{f,g}$, 112
$e_{A,g}(t,t_0)$, 202
$e_{f,g}(t,t_0)$, 105
$f(t_0+)$, 2
$f(t_0-)$, 2
g-derivative, 60
g-differentiable function, 60
kth Stieltjes derivative, 60

A
asymptotically stable solution, 272

B
Banach indicatrix, 38

C
Cauchy mean value theorem, 74
center, 314
characteristic equation, 157

continuous component, 78
continuous matrix, 194
critical point for Stieltjes differential system, 300

D
decreasing function, 1
Dirichlet–Stieltjes boundary conditions, 336

E
equilibrium point for Stieltjes differential system, 300
exponents, 263

F
Fermat theorem, 71
first Stieltjes boundary conditions, 336
Floquet multipler, 263
formula for integration by parts, 44
function of exponential order, 119
function of finite variation, 17
fundamental matrix, 216
fundamental set of solutions, 153, 216
fundamental system of solutions, 153

G
general solution, 216

H

homogeneous second-order linear Stieltjes differential equations, 152
homogeneous Stieltjes boundary value problem, 336

I

increasing function, 1
isolated critical point for Stieltjes differential system, 301
isolated equilibrium point for Stieltjes differential system, 301
isolated rest point for Stieltjes differential system, 301
isolated singular point for Stieltjes differential system, 301
isolated stationary point for Stieltjes differential system, 301

L

left saltus of a function at a point, 2
Lipschitz condition, 18

M

Monodromy matrix, 263
monotonic function, 1

N

nonhomogeneous second-order linear Stieltjes differential equation, 152
nonhomogeneous Stieltjes boundary value problem, 336

O

orbit of Stieltjes differential system, 298

P

path of Stieltjes differential system, 298
periodic Stieltjes boundary conditions, 337
phase plane, 298
phase portrait of Stieltjes differential system, 302

Q

quasi-linear Stieltjes systems, 290

R

regular Stieltjes boundary value problem, 337
rest point for Stieltjes differential system, 300
restrictively stable system, 288
right saltus of a function at a point, 2
Rolle theorem, 73

S

saddle point, 308
saltus function, 5
saltus of a function at a point, 2
second Stieltjes boundary conditions, 336
separated Stieltjes boundary conditions, 336
singular point for Stieltjes differential system, 300
singular Stieltjes boundary value problem, 337
solution of SBVP, 338
stable focus, 314
stable improper node, 310
stable node, 304
stable proper (star-shaped) node, 309
stable solution, 272
stable system, 287
stationary point for Stieltjes differential system, 300
Stieltjes adjoint differential system, 228
Stieltjes circle square, 101
Stieltjes derivative, 60
Stieltjes differentiable function, 60
Stieltjes exponential function, 105
Stieltjes integral, 41

Stieltjes matrix exponential function, 202
Steiltjes primitive, 79
Stieltjes regressive function, 91
Stieltjes regressive group, 95
Stieltjes regressive matrix, 195
Stieltjes regressive second-order Stieltjes differential equation, 153
Stieltjes transitive matrix, 221
Stieltjes–Euler–Cauchy equation, 163
Stieltjes–Laplace transform, 121
Stieltjes–Liouville formula, 207
Stieltjes–Putzer algorithm, 240
Stieltjes–Wronskian, 153
strictly decreasing function, 1
strictly increasing function, 1
strictly monotonic function, 1
superposition principle, 215

T

Taylor polynomial, 88
total variation, 17
trajectory of Stieltjes differential system, 298

U

uniformly stable solution, 281
unstable focus, 314
unstable improper node, 310
unstable node, 305
unstable proper (star-shaped) node, 310
unstable solution, 272

www.ingramcontent.com/pod-product-compliance
Lightning Source LLC
LaVergne TN
LVHW022314291224
800089LV00002B/53